# The Emergence of China's Smart State

# DIGITAL TECHNOLOGIES AND GLOBAL POLITICS

Series Editors: Andrea Calderaro and Madeline Carr

While other disciplines like law, sociology and computer science have engaged closely with the Information Age, international relations scholars have yet to bring the full analytic power of their discipline to developing our understanding of what new digital technologies mean for concepts like war, peace, security, cooperation, human rights, equity, and power. This series brings together the latest research from international relations scholars—particularly those working across disciplines—to challenge and extend our understanding of world politics in the Information Age.

*Governing Cyberspace: Behaviour, Power and Diplomacy* edited by Dennis Broeders and Bibi van den Berg
*Internet Diplomacy: Shaping the Global Politics of Cyberspace* edited by Meryem Marzouki and Andrea Calderaro
*The New Knowledge: Information, Data and the Remaking of Global Power* by Blayne Haggart and Natasha Tusikov
*Hybridity, Conflict, and the Global Politics of Cybersecurity* edited by Fabio Cristiano and Bibi van den Berg
*The Emergence of China's Smart State* edited by Rogier Creemers, Straton Papagianneas, and Adam Knight

# The Emergence of China's Smart State

Edited by Rogier Creemers, Straton Papagianneas, and Adam Knight

ROWMAN & LITTLEFIELD
*Lanham • Boulder • New York • London*

Credits and acknowledgments for material borrowed from other sources, and reproduced with permission, appear on the appropriate pages within the text.
Published by Rowman & Littlefield
An imprint of The Rowman & Littlefield Publishing Group, Inc.
4501 Forbes Boulevard, Suite 200, Lanham, Maryland 20706
www.rowman.com

86-90 Paul Street, London EC2A 4NE



British Library Cataloguing in Publication Information Available

Library of Congress Cataloging-in-Publication Data Available

978-1-5381-8441-7 (cloth : alk. paper)

978-1-5381-8442-4 (ebook)

# Contents

# List of Abbreviations

| | |
|---|---|
| 3GPP | Third Generation Partnership Project |
| AMS | Academy of Military Science |
| BRI | Belt-Road Initiative |
| CAC | Cyberspace Administration of China |
| CBDC | Central Bank Digital Currency |
| CCP | Chinese Communist Party |
| CCPCC | Central Committee of the Chinese Communist Party |
| CCIC | Central Cybersecurity and Informatization Commission |
| CII | Critical Information Infrastructure |
| CIIF | China Internet Investment Fund |
| CLGCI | Central Leading Group for Cybersecurity and Informatization |
| CNCERT/CC | National Computer Network Emergency Response Technical Team/Coordination Center of China |
| CNNIC | China Internet Network Information Center |
| CPD | Central Propaganda Department |
| CSL | Cybersecurity Law |
| DDoS | Distributed Denial of Service |
| DNSO | Domain Name Supporting Organisation |
| DoJ | Department of Justice |
| DSL | Data Security Law |
| DSR | Digital Silk Road |
| EDA | electronic design automation |
| EU | European Union |
| FRAND | fair, reasonable, and non-discriminatory terms |
| FT | Financial Times |
| FTA | free trade agreement |
| FYP | five-year plan |
| GATS | General Agreement on Trade in Services |

| | |
|---|---|
| GVC | global value chain |
| GBA | Greater Bay Area |
| GGI | Group of Government Experts |
| GPA | Agreement on Government Procurement |
| GDSI | Global Data Security Initiative |
| ICANN | Internet Corporation for Assigned Names and Numbers |
| ICT | information and communication technology |
| IEC | International Electrotechnical Commission |
| IETF | Internet Engineering Task Force |
| IP | Intellectual Property |
| ISA | Instruction Set Architecture |
| ISO | International Standardisation Organisation |
| ITU | International Telecommunication Union |
| IW | Information Warfare |
| MFA | Ministry of foreign Affairs |
| MIIT | Ministry of Industry and Information Technology |
| MOFCOM | Ministry of Commerce |
| MoST | Ministry of Science and Technology |
| MPS | Ministry of Public Security |
| MSS | Ministry of State Security |
| NDRC | National Development and Reform Commission |
| NPC | National People's Congress |
| O2O | Online-to-offline |
| OEWG | Open-Ended Working Group |
| PBoC | People's Bank of China |
| PIPL | Personal Information Protection Law |
| PLA | People's Liberation Army |
| PRC | People's Republic of China |
| SPC | Supreme People's Court |
| SCR | Smart Court Reform |
| SCS | Social Credit System |
| SDO | standard development organisation |
| SEP | standard-essential patent |
| SEZ | Special Economic Zone |
| SIIO | State Internet Information Office |
| SPS | Agreement on Sanitary and Phytosanitary Measures |
| SME | small and mid-size enterprises |
| SSF | Strategic Support Force |
| SWIFT | Society for Worldwide Interbank Financial Telecommunications |

| | |
|---|---|
| TBT | Technical Barriers to Trade |
| TC260 | Technical Committee 260 |
| TPP | Third-party |
| UN | United Nations |
| US | United States |
| W3C | World Wide Web Consortium |
| WGIG | World Group on Internet Governance |
| WIC | World Internet Conference |
| WLAN | Wireless Local Area Networking |
| WSIS | World Summit on the Information Society |
| WTO | World Trade Organisation |

# Introduction

## Rogier Creemers

In February 2014, the Chinese government established a new, top-level coordinating body for the digital realm, the Central Leading Group for Cybersecurity and Informatisation. On this occasion, Chinese Communist Party (CCP) General Secretary Xi Jinping, the chair of this new entity, outlined a simple but ambitious aspiration. China had become a "large network country" (*wangluo daguo*) but now should develop into a "strong network country" (*wangluo qiangguo*). This entailed that China should generate world-class indigenous technologies, top-notch information services, flourishing online culture, solid network infrastructure, and a powerful digital economy. This all should be supported by increasingly capable scientists and engineers, and strongly improved research and development (China Copyright and Media 2014).

In pursuit of this strategy, the leadership ordered a complete overhaul of the digital governance architecture, establishing several new bodies and reorganising lines of authority of existing ones (Creemers 2020). A whole range of policy plans emerged, with the 13th Five-Year Planning cycle in 2016. For the first time, a dedicated document was devoted to national informatisation, listing seventy-four discrete issue areas ranging from expanding cloud computing centres and 5G access to expanding funding channels for higher education in STEM (State Council 2016). Specific plans were drafted for critical areas including artificial intelligence, big data and "Internet Plus," or the integration of digital capabilities with legacy manufacturing activities. Major policy initiatives on "novel infrastructure," smart cities, and the application of blockchain technologies in non-fintech areas not only intend to facilitate growth, but to increase the quality of life and access to public services of ordinary Chinese individuals. The urban-rural divide is being targeted through projects involving long-distance education and healthcare. Legislative initiatives produced the Cybersecurity Law, the Personal Information Protection Law, the Data Security Law, and a host of subordinate regulations that provided more clarity on the leadership's goals and aims and sought to build the protective exoskeleton needed in a connected society. A national cybersecurity centre was established in Wuhan, with the purpose of training half a million cybersecurity professionals over the next decade (Cary 2021). New methods for financing these projects came into being, ranging from local government guided investment funds (Pan et al. 2021), the Shanghai STAR stock exchange market (Lu and Ye 2019) and the "Big Fund" (Li 2021), which has mobilised over 500 billion RMB for investment in the semiconductor industry. Naughton has described the combination of prioritising high-tech development and informatisation with using market-oriented mechanisms such as investment funds, subsidies, and tax breaks as "grand steerage," a part of a government drive to search for new sources of economic growth. These efforts not only took place at the central level. Local governments, too, have responded vigorously to the call to develop technological capabilities.

Yet it is one thing to impose a "top-level design" (*dingceng sheji*) as an expression of political will and virility, it is another to navigate the many complexities that beset digital development. The grand steerage approach carries significant risks of duplicate or excessive investment, destroying economic value. The large amounts of money sloshing around in these investment funds make for tempting opportunities for corruption. In the summer of 2022, multiple senior semiconductor executives and officials came under investigation, including Minister of Industry and Information Technology Xiao Yaqing (White and Liu 2022). A wave of regulatory action was required to bring China's large platform companies under control, among others to maintain stability in the financial system and rectify market failures arising from their dominant positions (Creemers 2023). The introduction of AI-enabled automated decision-making processes in judicial and regulatory entities has run into headwinds over both technical and political difficulties. Most damagingly, however, was the pushback coming from abroad. China's digital ambitions and achievements were at the core of greater tensions between Beijing and Washington, which have translated into ever broader and more impactful sanctions. These started with companies such as Huawei and ZTE being banned from some Western markets in 2018 and have reached a high water mark with the blanket prohibition of cooperating with Chinese companies in the production of high-end semiconductors in October 2022. The impact of these sanctions is severe: Huawei briefly was the largest smartphone manufacturer in the world but has now dropped out of the global top five. The semiconductor sanctions are intended to consign China to permanent second-class status in this core area of technology, by severing the networks of financial, technological, and personnel exchange between China and the United States. Moreover, concerns about China's growing international footprint have led to its facing ever more resistance in its bid to become more influential in setting technical standards, as well as in the establishment of norms for governing the Internet itself, or the conduct of states in the digital realm. Both in China and elsewhere in the world, globalisation choices are being reassessed as the increasing territorialisation of the digital realm imperils cross-border data flows, and information exchange taken for granted hitherto.

These evolutions raise diverse and complex empirical, conceptual, and theoretical questions. How does the overall, abstract ambition to become a digital power, or to "smarten" the state, translate into specific goals, and achievable outcomes? How do state institutions reform to meet the challenges this entails? Which tactics and techniques are used within specific sectors or in particular localities to achieve these goals? How do changing domestic and international circumstances facilitate or challenge the achievement of these objectives? How do different interest groups and policy considerations interact? And how are compromises or trade-offs between them navigated? Yet the academic importance and real-world impact of these topics notwithstanding, preciously little academic attention has been devoted to them. Nearly a decade after its foundation, for instance, not a single publication addresses the functioning of the Cyberspace Administration of China (CAC), arguably the most powerful Internet governance body in the world. China's progress in strategic emerging technologies, as well as in expanding its international influence, remain blind spots in the scholarly literature.

The goal of this book is to understand and explain the various facets of China's digital ambitions and the policies by which it seeks to realise it. That means this is a book, first and foremost, about China itself. That means two things: first, it does not primarily approach the subject of technology in China from the angle of Sino-American tensions, a dominant theme in the literature at present. Instead, the goal of this book is to adopt a Beijing-centric perspective. Second, this book does not take an evaluative approach that attempts to gauge the extent to which China meets the criteria of any particular academic or normative framework. As Don Clarke (2003) already warned two decades ago, doing so actually tells us very little about what animates or informs decision-making and policy evolutions in China. It also often leads to us considering China to be an aberration and blinds us to perceiving the logic and rationality of the Chinese system on its own terms. Lastly, the point of this book is not to declare victory or defeat of the digital power strategy or smart state ambitions. Rather, it is to highlight the dynamic changes, complexities, and contradictions inherent in China's digital development policies.

This book brings together authoritative voices from the academic and think tank world to open up these debates across four topic areas. The first is conceptual, discussing how China conceives of the role of digital technologies in its development process, and how institutional reforms are made to realise these. The second addresses China's progress in certain strategic emerging technologies, including fintech, semiconductor manufacturing, and blockchain-enabled services. The third addresses the local component of the digital power strategy, reviewing how local governments have responded to the gradual expansion of digital ambitions. The last reviews cross-border processes and China's engagement with global technical governance processes.

## REFERENCES

Cary, Dakota. 2021. China's National Cybersecurity Center: A Base for Military-Civil Fusion in the Cyber Domain, CSET, July. Available from: https://cset.georgetown.edu/publication/chinas-national-cybersecurity-center.

China Copyright and Media. 2014. Central Leading Group for Internet Security and Informatization Established. 1 March. Available from: https://chinacopyrightandmedia.wordpress.com/2014/03/01/central-leading-group-for-internet-security-and-informatization-established.

Clarke, Donald. 2003. Puzzling Observations in Chinese Law: When Is a Riddle Just a Mistake? In Hsu, C. Stephen, ed. Understanding China's Legal System. New York University Press, 93–121.

Creemers, Rogier. 2020. China's Cyber Governance Institutions. Leiden Asia Centre Report. Available from: https://leidenasiacentre.nl/report-chinas-cyber-governance-institutions.

Li, Yin. 2021. The Semiconductor Industry: A Strategic Look at China's Supply Chain. In Spigarelli, Francesca, and John McIntrye, eds. The New Chinese Dream. Palgrave Macmillan, 121–36.

Lu, Lerong and Ningyao Ye. 2019. Promoting High-tech Innovations through Capital Markets Law Reform: Deciphering the Sci-Tech Innovation Board of the Shanghai Stock Exchange. *Journal of International Banking and Financial Law* 35: 140–43.

Naughton, Barry. 2022. Grand Steerage as the New Paradigm for State-Economy Relations. In Pieke, Frank, and Bert Hofman, eds. CPC Futures: The New Era of Socialism with Chinese Characteristics. NUS East Asian Institute, 105–12.

Pan, Fenghua, Fangzhu Zhang, and Fulong Wu. 2021. State-led Financialization in China: The Case of the Government-guided Investment Fund. The China Quarterly 247: 749–72.

State Council. 2016. "Shisan wu" guojia xinxihua guihua (13th Five-Year Plan for National Informatization), issued 15 December. Available from: http://www.gov.cn/zhengce/content/2016-12/27/content_5153411.htm.

White, Edward and Qian'er Lu. 2022. China's Big Fund Corruption Probe Casts Shadow over Chip Sector. Financial Times, 29 September. Available from: https://www.ft.com/content/8358e81b-f4e7-4bad-bc08-19a77035e1b4.

# PART I

# Digital Concepts and Institutions

# *Chapter 1*

# The Cyberspace Administration of China

## *A Portrait*

Jamie Horsley and Rogier Creemers

### INTRODUCTION

Within the space of a short few years, the Cyberspace Administration of China (CAC) has become arguably the most powerful digital institution in the world. It lies at the centre of a new institutional architecture whose task is to build China into a "cyber superpower" (*wangluo qiangguo* 网络强国) and realise the ambitious "cybersecurity and informatisation" agenda, but it defies easy description. It operates within both the Chinese Communist Party (CCP) and the state hierarchies and, while it has clear regulatory tasks, it is not merely a regulator. It coordinates the implementation of policies of the Party Central Cybersecurity and Informatisation Commission[1] (CCIC), reportedly composed of the heads of all digital-relevant Party and State bodies, but it is not an interagency body itself. Its tasks range from overseeing the vast Chinese online censorship apparatus to promoting the expansion of connectivity in rural regions, from highly technical tasks such as regulating algorithms, overseeing cybersecurity standards, and managing cyber incident response, to more political and diplomatic ones such as international outreach. Bureaucratically, it is a sui generis entity that does not neatly fit existing categories of ordinary Party or State bodies. Moreover, its early history is overshadowed by the towering ambitions and subsequent fall of its first independent director, Lu Wei.

The CAC inevitably must navigate many of the contests and contradictions characterising Chinese digital politics. On the one hand, China intends to become a technological leader with greater international competitiveness and self-sufficiency, universal high-bandwidth connectivity and a powerful digital economy. On the other, Beijing has just conducted a wide-ranging regulatory offensive to reshape the online platform economy, it maintains perhaps the most elaborate content censorship system in the world, and it is imposing ever stricter regulations surrounding data protection and cybersecurity. Surprisingly, in view of the importance of Chinese digital policies for China's overall future as well as global cyber affairs, and the centrality of the CAC's role in it, little academic and analytical attention has, thus far, been devoted to this institution.

This chapter will explore the multiple faces of the CAC, providing a portrait of the institution in the breadth of its roles. First, it will discuss its historical development, describing its rapid emergence from obscurity to become a leading player in the digital space by 2014, as well as the personalities and political dynamics of that period. A second section will focus on its institutional composition, the scope of its powers and responsibilities, as well as the mechanisms by which it oversees its local subordinates and the specialised technical bodies over which it has authority. A third section will discuss the CAC's nature and functioning within the broader Chinese bureaucratic landscape. The final section identifies emerging questions about the efficacy and strength of the CAC and, more broadly, Chinese digital policies in the light of the ambitions outlined in the recent 14th Five-Year Plan.

## THE HISTORICAL DEVELOPMENT OF THE CAC

The CAC was founded in May 2011, under the name of the State Internet Information Office (SIIO). At that time, it was a relatively unobtrusive department of the State Council Information Office, itself the State face of the Party Central Propaganda Department (CPD), without independent staffing. The litany of problems it was intended to address remain well-recognised, including "false information and malicious speculation, pornographic and vulgar information, fraud and gambling, illegal marketing, etc.." Moreover, presaging Xi Jinping's dialectic view of cybersecurity and informatisation, the vision that the SIIO was set out to realise in the online information space was that "development and management complement each other, development requires management, and management enables sound and fast development." The list of responsibilities this new department should bear was, however, rather out of kilter with its relatively low bureaucratic status and shortage of resources, including licensing of online businesses, overseeing online games, video and publishing, managing online news and propaganda, law enforcement against illegal websites, and coordinating the management of telecommunications and internet access providers, the domain name system and other elements of infrastructure (Xinhua 2011).

Very rapidly, however, that started to change. The establishment of the SIIO reflected a growing awareness in the central leadership that the Internet, and particularly smartphone-driven user-generated content, was starting to have a transformational impact on information circulation. This became particularly apparent through a raft of political scandals emerging on social media platform Weibo (Wright 2017). At the same time, other security concerns gained prominence, highlighted by incidents such as the Snowden revelations and Microsoft's announced cessation of security support for Windows XP at a time when over two thirds of Chinese computers still used that system (Creemers 2017). Following Xi Jinping's accession to the CCP General Secretaryship in 2012, the SIIO's profile rose rapidly. This happened under the leadership of Lu Wei, the previous propaganda chief in the Beijing Municipal Government who was appointed as SIIO's first independent director. Lu had already made himself a reputation in countering the raucous social media sphere, developing new approaches to deal with the "big V" online celebrities. He immediately started turning the SIIO into a very visible force against undesirable online information. This would result, in the summer of 2014, in its gaining explicit responsibility for the governance of all online content (State Council 2014).

In the meantime, a bigger change had taken place as well. To integrate digital policy and increase its political visibility, a new top-level coordinating body, the Central Leading Group for Cybersecurity and Informatisation (CLGCI) had been established in 2014. Xi Jinping chaired this group, with State Council Premier Li Keqiang and previous propaganda chief Liu Yunshan as deputies. Secretarial responsibilities were given to SIIO, which now took the additional title of the Office of the CLGCI. In English, it started using the name Cyberspace Administration of China in the summer of 2014, while its Mandarin name remained unchanged. As a result, CAC's actual bureaucratic nature remains vague. As the SIIO, it appeared to be subordinate to the State Council, but as the Office of the CLGCI, it was a Party institution. As such, its *bianzhi*, the official inventory of its responsibilities, internal structure, and staffing numbers, has not been made public. The ramifications of this dual Party-state structure will be discussed in depth below.

With this new elevation came expansion. Soon after the establishment of the CLGCI, two departments of the Ministry of Industry and Information Technology (MIIT), together with their staff, were transferred to CAC (CIOC 2015). One of these was the Cybersecurity Coordination Department, which holds certain authority over Technical Committee 260 (TC260), the technical cybersecurity standard-setting body. CAC also gained enforcement competences, with the transfer of the China Internet Unlawful and Harmful

Information Reporting Centre from the Internet Society of China (CIUHIRC 2014). In December 2014, CAC took authority over the China Internet Network Information Centre (CNNIC), whose responsibilities include acting as China's domain name registry (Guangming Daily 2014). A charitable foundation for Internet development and an in-house think tank, the China Academy of Cyberspace Studies, were established in 2015, followed the year after by the China Internet Investment Fund (CIIF), managed together with the Ministry of Finance (CAC 2019). This fund holds ownership stakes in businesses including ByteDance, Weibo, and SenseTime (Economist 2021). In 2018, CAC also gained authority over the National Computer Network Emergency Response Technical Team/Coordination Center of China (CNCERT/CC), the entity that is also assumed to run the technical side of the Great Firewall (Marczak et al. 2015). This transfer was part of a broader reorganisation (Central Committee 2018) that saw the CLGCI upgraded to a Central Commission status, directly under the Party Central Committee, although there are no clear indications that its previous membership or mandate have undergone significant alterations.

The flamboyant Lu Wei also sought to build his international profile. In June 2014, he made his first high-profile foreign appearance, delivering a keynote speech at the 50th ICANN meeting in London (Lu 2014). He also visited the United States in that year, chairing the Sino-US Internet Forum and cultivating warm relationships with several technology executives. Mark Zuckerberg, eager to gain access to the Chinese market, welcomed him to Facebook's headquarters (Kan 2014). His initiative with the most durable legacy is the World Internet Conference, organised in the Zhejiang town of Wuzhen. This gathering intended to echo the London Process and NetMundial and grow China's "discursive power" (*huayuquan* 话语权) in the burgeoning debates on global digital governance. Organised annually since 2014, it was reformed into an "international organisation" in 2022 (Xinhua 2022).

Lu's flamboyance and empire-building had, however, gained him powerful enemies within the system, and he left his position in June 2016. He would later be sentenced to fourteen years in prison on corruption and malpractice charges (Gao 2019). Lu's successors, Xu Lin and later Zhuang Rongwen, were both far calmer apparatchiks, tasked with turning CAC into a coordinating body that could cooperate effectively with the State Council ministries involved in digital policy, as well as a competent executive department. While interdepartmental relationships and tensions are difficult to gauge from outside the system, there are indications that CAC has come to work more closely together with its counterparts. Regulation, enforcement, and management have become multientity tasks, as detailed below. At the same time, the multidepartmental regulatory environment has allowed CAC new opportunities to consolidate its position: as the primary drafting body for major documents including draft State Council regulations implementing the Cybersecurity Law (CSL), the Data Security Law (DSL) and Personal Information Protection Law (PIPL) (NPC 2016, 2021, 2021a), all assigning substantial regulatory powers to CAC (Creemers 2022), its position and authority seem assured.

## THE CAC'S POWERS, RESPONSIBILITIES, AND ACTIVITIES

As indicated earlier, no enumerated list of CAC competences currently exists, and thus its powers and responsibilities need to be inferred from broad legislative and policy frameworks, media reports, and its own statements and issued documents. Here, a measure of fuzziness is inevitable: CAC might formulate a document because it *has* responsibility in a certain policy area, or sometimes, because it *claims* it. Furthermore, CAC is more than purely a regulator. It not only sets, implements, and enforces rules for the conduct of businesses and individuals in digital processes but also plays important roles in supporting the work of the CCIC to realise China's ambitious "informatisation agenda." It holds authority over specialised technical bodies as well as the sector-specific business associations that form the interface between the state and private industry. It engages in international outreach and is even a shareholder in several Chinese tech firms through the CIIF. In short, CAC's mission concerns, on its own or in coordination with other bodies, the overall governance of the Chinese digital realm in line with overall national policy goals.

### Regulatory Responsibilities

#### *Online Content*

Online content management, assigned to CAC by the State Council in 2014, is arguably the area where it has been the most successful and effective. This is unsurprising: CAC has inherited and continued content control practices going back decades, and the necessity of censorship and content management are not politically contentious. CAC did bring a new approach to the online environment. Hitherto, legacy regulators had governed online media like traditional media, relying on tools such as prepublication review that were increasingly obsolescent in times of social media and user-generated content. CAC, in contrast, outsourced management tasks to platform companies themselves, culminating in the concept of "principal responsibility" (*zhuti zeren* 主体责任). This makes companies liable for the legality of content posted on their platforms, requiring them to dedicate sufficient human, financial, and technological resources to monitor content adequately, and respond to reports, complaints, and incidents (CAC 2021).

Initially, the CAC acted principally on the basis of the State Council's 2014 authorisation decision to regulate online information content (for instance, CAC 2015). The CSL changed this, including explicit provisions identifying undesirable content as a cybersecurity concern, and appointing CAC in charge of coordinating all cybersecurity-related work. Subsequently, CAC has issued regulations on content and related services including, but not limited to, search engines, mobile apps, livestreaming, forums, posting and commenting functions, public social media accounts, community information services, online news services, microblogs, blockchain information services, live-streamed marketing, financial information services, religious information services, algorithms, and AI-generated "deep synthesis" content (Creemers 2022). In other words, CAC not only regulates online information content *per se* but also the means to produce and distribute it.

*Personal Information Protection and Data Security*

The CSL only contained very basic provisions on data protection, and, apart from its overall coordination role, only explicitly empowered the CAC to oversee the export of personal information. This lack of clarity led to rivalry between the CAC and the Ministry of Public Security (MPS) as they both sought to establish authority over personal information protection and the emerging field of data security. Both institutions issued several draft regulations addressing elements such as data protection in critical infrastructure and data export, as well as general implementing regulations for personal information and data security. Most of these, however, were never adopted and implemented. The sole exception was a CAC document on the personal information protection of minors (CAC 2019), as child protection became a major overall priority of the leadership (Daum 2022). The regulatory conflict was only resolved with the promulgation of the PIPL and the DSL (Creemers 2022a). The former assigned overall responsibility for personal information protection work to CAC. The latter indicated that CAC would be responsible for "overall coordination" of data security regulation, albeit under the leadership of the CCP's National Security Commission. Since then, CAC has issued a draft of general implementing rules covering the CSL, DSL and PIPL (CAC 2021a) and definitive data export rules (CAC 2022), as well as detailed rules for the automobile industry (CAC 2021b). More sector-specific documents are likely on the way.

### *Cybersecurity*

The CSL gave CAC responsibility for overall planning, coordination, and supervision and management of cybersecurity work. It also established several specific technical mandates for CAC, including the protection of critical information infrastructure (CII), cybersecurity review of online products and services, and cybersecurity incident response. The first of these mandates again caused friction with the MPS, as there were clear overlaps between the mooted CII regime and the already existing Multi-Level Protection System (MLPS) run by the MPS. This imposes incremental security requirements on network operators, depending on the importance of their systems. Resolution of this overlap came in 2021, when the State Council issued comprehensive regulations on CII protection, giving a coordinating role to CAC and an executive function to MPS (State Council 2021).

CAC also holds responsibility for "cybersecurity review" (*wangluo anquan shencha*). Originally, because of 2017 rules under the CSL, this was a relatively limited mandate, focusing predominantly on assessing whether particular online products could be securely included into telecommunications networks (CAC 2017). At that time, CAC also established a Cybersecurity Review Committee, which has since been upgraded into the Cybersecurity Review Office. This Office has gained quite some notoriety in recent years, however, imposing cybersecurity reviews on, among others, ride hailing giant Didi after its IPO on the New York Stock Exchange in 2021, and on academic database operator CNKI in 2022 (DigiChina 2022a, Chen and Bandurski 2022). Both these reviews were broadly seen as having political grounds, rather than technical cybersecurity concerns. Revised cybersecurity review measures reflected this shift, expanding the grounds for review to include nontechnical elements such as foreign listings of businesses holding large amounts of data on Chinese citizens and compliance with DSL requirements (CAC at al. 2021). This effectively turns CAC into a securities regulator of sorts and creates new questions about its interaction with legacy financial and securities authorities.

A last important CAC task is cyber incident response. This is where CNCERT/CC plays an important technical role in countering the sources of an attack, but the CSL's conception of cybersecurity is far broader than hacks or intrusions alone. Consequently, CAC has issued an overall plan that defines incidents into categories including malware, attacks, equipment failures, natural disasters as well as information security incidents. It also established a National Cybersecurity Emergency Office that continuously monitors the cybersecurity status and maintains a readiness state consisting of blue, yellow, orange, and red tiers. These, in turn, affect the level to which other government departments must maintain the state of alert of their resources and personnel (CAC 2017a)

In these different areas, CAC's authority is rarely completely exclusive. Even in content, traditional media regulators such as the National Radio and Television Administration retain powers over, for instance, the production of audio-visual programmes, even if their distribution primarily takes place online. In other areas too, CAC collaborates with a range of line ministries in passing, implementing and enforcing regulations, a sign of both its growing maturity as a coordinating body and the nature of the governance questions it is tasked to address: as "the digital" penetrates into ever more aspects of daily life, digital policy will become less of a discrete regulatory sphere.

### *Policy Coordination*

In its role as Office of the CCIC, CAC has prime responsibility in coordinating the drafting and promulgation of overall policy documents pertaining to digital policy. To be sure, none of the three major documents in the 14th Five-Year Plan cycle dealing with the digital realm was published under CAC's name. Instead, the overall plan was issued by the CCIC itself (CCIC 2021), with the more detailed documents for digital government and the digital economy published respectively by the National Development and Reform Commission, and the State Council. Still, it can be expected that the CAC had considerable input, most notably at the level of the CCIC. CAC also convenes deliberation events at the working level, such as a recent "National Online Civilisation Construction Work Progress Meeting" attended by representatives from fifty-seven member entities of the CCIC and the Central Civilisation Committee, as well as provincial representatives (CAC 2022). Lastly, CAC has a particular role in coordinating different agencies in furthering the advance of connectivity, publishing regular plans including, since 2020, annual plans for developing the "digital countryside" (CAC 2020).

The CAC also oversees the activities of subordinate, technically specialised bodies. In most cases, that line of authority is direct, with CAC having either established them itself, as in the case of the Chinese Academy of Cyberspace Studies, or having its authority recognised by official writ, such as with CNNIC. One exception is TC260, the National Information Security Standardisation Technical Committee nominally affiliated with MIIT. This has been chaired, for many years, by Zhao Zeliang, whose main function is CAC Chief Engineer and Deputy Director, and was previously the head of its Cybersecurity Coordination Bureau (CAC undated). On top of this personnel linkage, the most authoritative policy document on cybersecurity standardisation was issued with CAC as lead entity (CAC, AQSIQ, and SSMC 2016) and CAC and TC260 regularly co-organise relevant events. Although there is no direct evidence of administrative oversight, it can be assumed that CAC can exert considerable influence on the functioning of TC260, even if the latter's membership is predominantly made up of either technical experts and scholars, or representatives from private businesses. CAC also supervises several industry associations, such as the Cybersecurity Association of China (CSAC 2019). These organisations act as an interface between the Party-state and the private sector, enabling communication in both directions. They also have "self-regulatory" roles, with their members committing to codes of conduct outlining not just best industry practices, but also politically expected behaviour. Another link between CAC and private industry is, sometimes, direct ownership through the CIIF, which may enable CAC to exercise more direct control over companies but certainly is a manner to obtain internal corporate information more easily.

### *The CAC as Cheerleader*

Outside of its regulatory and policy responsibilities, the CAC often acts as a cheerleader for the expanding connectivity of Chinese society, and the Centre's digital agenda more broadly, and in setting the tone of public discourse. Sometimes, these efforts may appear rather campy to Western observers, such as the performance of the rousing patriotic anthem "The Spirit of Cyberspace" at a televised Lunar New Year performance in 2015 (Mozur 2015). Its lyrics included "Unified with the strength of all living things, devoted to turning the global village into the most beautiful scene" and "An Internet power: Tell the world that the Chinese Dream is uplifting China." Somewhat more seriously, Lu Wei adopted the phrase "positive energy" (Lu 2013), which emerged in documents on propaganda around the time Xi Jinping came to power. This has become a core term in digital culture, with CAC posting a list of five hundred recommended pieces of positive energy content it deemed sufficiently imbued with the correct Party values (Boyd 2022). Conversely, it also tackled fake news, launching the "China Internet Joint Rumour Countering Platform" (Zhongguo hulianwang lianhe piyao pingtai) in 2018. On this website, CAC and State Council ministries provide official refutations of supposed online rumours, thus acting as antimisinformation fact checkers. Illustratively, it published a top-ten list of rumours dubbed "historical nihilism," the general denomination for historical accounts deviating from the officially adopted one (Boyd 2021).

Not all CAC propaganda is ideological in nature. It also aims to foster cybersecurity awareness and propagate China's triumphs in its technological advance. Since 2014, CAC has organised an annual National Cybersecurity Week. This encompasses a series of events, exhibitions, radio and television programmes, academic and policy exchanges, co-organised with other ministries or private companies, to raise cybersecurity awareness among the general population. Much of the Wuzhen World Internet Conference serves as a showcase for the prowess of Chinese technology companies, for instance through the "Light of the Internet" exhibition taking place every year (WIC undated). More broadly, the CAC has propagated the role of technology in broader Party initiatives, such as the drive to eliminate extreme poverty (Xinhua 2020).

### *Foreign Engagement*

In its early days, under Lu Wei, CAC not only attempted to make its mark domestically, it also sought to establish itself as the primary Chinese body engaging with global digital commerce and governance processes. As indicated earlier, Lu Wei racked up several high-profile appearances at events like ICANN's London meeting and the China-US Internet industry Forum. Furthermore, CAC sent its own delegations to international events such as the 2015 Global Conference on Cyberspace, where CAC Deputy Director Wang Xiujun gave a speech (Xinhua 2015). Following Lu Wei's departure, however, CAC's role in foreign affairs has diminished considerably. It still participates in international diplomatic and publicity-oriented engagements, but with the Ministry of Foreign Affairs (MFA) in the lead. For instance, CAC has promoted Xi's vision of jointly building a community with a shared future in cyberspace with Africa (CAC 2021c) and sponsored an APEC symposium on using digital technology for poverty reduction (CAC 2021d) together with the MFA. It has also lobbied for international support of China's Global Initiative for Data Security (MFA 2020), a proposed model for handling data storage and digital commerce security (DigiChina 2022; Webster and Triolo 2020).

CAC's most notable effort to connect with the outside world remains the World Internet Conference (WIC), organised annually in the Zhejiang river town of Wuzhen since 2014. Within China, this event has become very prestigious. A completely new conference centre was built for it, and Politburo Standing Committee members routinely speak at its opening ceremony. In 2015, Xi Jinping personally attended, delivering a speech that still forms the foundation of China's approach to digital diplomacy (Xi 2015). In terms of gaining international traction, however, Wuzhen has had little impact. At the WIC's first iteration, consternation arose among foreign attendees as a proposed 'Wuzhen Declaration' was posted under the doors of their hotel rooms, to be announced as reflecting their support for, essentially, China's approach to global Internet governance (Shu 2014). After heated arguments during the night, the draft declaration was, eventually, not released but did cause lingering mistrust about Chinese tactics. A subsequent 'Wuzhen Initiative,' presented as the product of the WIC's High-Level Advisory Council (WIC 2015), has equally not been taken up outside of China. A few years later, another initiative for digital economic cooperation and regulatory harmonisation only gained support from six other countries (WIC 2017). Overall, the WIC has never been as global as its name suggests, with foreign guests comprising a small minority of total attendance, and participation granted on an invitation-only basis with opaque conditions. The COVID-19 pandemic and the associated travel restrictions further limited foreign presence. Nonetheless, in July 2022, CAC announced the establishment of an opaque World Internet Conference International Organisation reportedly comprised of Internet-related organisations, enterprises, experts and scholars, as a platform to contribute to global Internet development and governance (WIC 2022). How this differs from the earlier WIC remains to be seen.

## THE CAC'S INSTITUTIONAL IDENTITY WITHIN CHINA'S BUREAUCRATIC LANDSCAPE

As outlined above, the CAC is charged by law with many responsibilities associated with government regulation, outside of its content control mandate. However, the CAC is not a traditional government agency. It is an opaque, seemingly dual Party-state institution, referred to as 'one institution, two nameplates.' This means that the "CAC" and the "Office of the CCIC" are two names of a single institution; the corresponding names can be used externally according to the party or state needs and nature of the work (CIOC 2014). As such, it is solely accountable to the CCP Central Committee, not the State Council, China's central government. Its original institutional parent, the State Council Information Office, is similarly a single institution with dual Party-state identities under the Central Committee, reporting to the Central Propaganda Department. CAC's current institutional parent is the CCIC, which as a deliberative and coordinating body under the Central Committee, takes its marching orders directly from the Party General Secretary Xi Jinping (Central Committee 2020).

Underlining the propaganda link, all CAC directors have concurrently held a deputy directorship of the CPD, which traditionally has overseen all party ideology and information dissemination and censorship work. Moreover, many—if not most—directors of provincial and lower cyberspace administrations (local CACs) are likewise concurrently deputy directors of the local propaganda department, and some local cyber authorities continue to be funded by the propaganda system. Even so CAC, in its party capacity, appears to now be of equal institutional stature with the CPD. Functionally, the CAC is arguably now the more powerful one of the two, as the Internet has become the dominant platform for information dissemination, and the CPD has become considerably less prominent in recent years.

CAC uses its Party identity as the CCIC Office when undertaking Party-related activities, such as policymaking and Internet-related cheerleading and propaganda initiatives. Little is known about how CAC interacts with the CCIC, whose membership, procedures, and meetings have not been disclosed. Unofficial reporting suggests the Commission includes minister-level officials from all Party and State bodies with a substantial stake in digital affairs, as well as the military (Guancha 2014). Reports on its work meetings have identified General Secretary Xi Jinping as the CCIC chairman, with Politburo Standing Committee Members Premier Li Keqiang, and ideological theorist Wang Huning as vice chairs (China Copyright and Media 2018). Other meeting reports provide the names and affiliations of some thirty attendees, but those appear to include members of both the CCIC and another body, the Central Civilisation Commission, so CCIC's exact membership is still not entirely clear (CAC 2022b). CAC uses its state identity, which in Mandarin remains the original "SIIO" (*guojia hulianwang xinxi bangongshi*) and includes the designation of "State," when conducting traditional government affairs. These include rulemaking (although it is not clear that its rules have the same status as government departmental rules), licensing and enforcement, which it appears to generally conduct in accordance with administrative law procedures that govern the State Council government agencies (Horsley 2022). For example, CAC typically publishes its draft rules for comment and incorporates input into the final version, which it generally files with the State Council for review and recording.

With cyberspace viewed as the main battleground of ideological struggle, impacting regime security as well as national security (Central Committee 2017), putting a directly led party institution in charge helps ensure the Party's leadership over the cyberspace domain. CAC's Party pedigree, combined with the range of regulatory responsibilities it has been assigned, would seem to provide it a status somewhat higher than the State regulators of the internet and informatisation sector, or at least on a par with the superministerial National Development and Reform Commission (NDRC), with all of which CAC must collaborate.

Paradoxically, to an extent not seen with other dual Party-state institutions, the Party has buttressed CAC's political power with State legal authority through laws adopted by the national legislature—the CSL, DSL, and PIPL—and nationwide regulations enacted by the State Council. Endowing CAC with a state aspect enhances its legitimacy as a regulator, even while it creates challenges for China's administrative law system (Lin 2019). Nonetheless, CAC's statutory role outside of its few clearly demarcated direct responsibilities, is generally framed in terms of coordinating and overseeing relevant

State departments such as MPS and MIIT as they carry out their respective duties. This is in line with general Central Committee instructions for the Party to support public and state security authorities to safeguard national security and investigate crimes and terrorist activities, and for industrial or sectoral ministries like MIIT to be primarily responsible for front line work, including cybersecurity inspections and handling cybersecurity incidents, while keeping the local CACs informed (Central Committee 2017).

As also required by the relevant law, CAC frequently collaborates with one or more State departments and Party institutions on policy documents, rulemaking, and enforcement actions, often but not always taking the lead when those impact the cybersecurity and informatisation sector. Jointly issued documents often call on different departments to implement them independently in accordance with their respective responsibilities. In other cases, CAC is supposed to share implementation responsibilities with State departments, such as through joint cybersecurity assessment review mechanisms housed within CAC's Cybersecurity Coordination Bureau (CAC 2021g). It also joined in 2019 with MIIT, MPS, and SAMR, as well as several associations and technical institutions, to form a working group to better regulate the illegal collection of personal information by online apps. The online app working group initiative included announcements, various measures, a new standard and publicised campaigns by different ministries and localities to crack down on app violations (Fang and Yu 2020). Also in 2019, the State Council tasked the NDRC to lead work with the CCIC Office, MIIT, SAMR and MPS to ensure sound development of the platform economy (State Council 2019). CAC, together with tax authorities, participated at least twice in interdepartmental regulatory guidance meetings led by SAMR with tens of platform companies to curb monopolistic and other unfair and illegal online conduct, including tax evasion and infringing personal information.

Possible bureaucratic tensions surface at times. CAC led an interagency drafting and issuance of cybersecurity review measures that stipulated general procedures in 2020. However, it acted more unilaterally in quickly publishing for comment and approving, with the "agreement" of twelve other regulators, the revision in 2021 that gave CAC new authority to review overseas listings by companies holding the personal information of one million or more users (CAC, etc. 2021) That revision provided retroactive authority for CAC's unexpected cybersecurity review of Didi and other Chinese platforms, accompanied by restrictions on related apps and new user registrations (Liu and Jia 2021). CAC's actions contributed to an extended regulatory onslaught that triggered market devaluations, employee lay-offs and foreign investor jitters. In March 2022, financial regulators appeared to push back, calling for greater regulatory coordination in announcing new policies that might impact the market (Xinhua 2022a). CAC subsequently assured the public of its support for Chinese companies to list overseas (Xinhua 2022b).

## Central-Local Relationships

CAC's own website links to thirty-one provincial CACs, and higher-level CACs are to supervise lower-level CAC work. However, institutionally, China's perennial problem of fragmented authority remains. Local CACs are established directly under the provincial party committees, with overlapping leadership including with the propaganda departments. CAC, in its guise as the Office of the CCIC, reported in August 2022 that all party committees at the levels of the center, provinces and municipalities had established cybersecurity and informatisation commissions to consolidate cyber ideology and security work under the Party's leadership (Xinhua 2022b). Lower-level CACs are similarly under the party committees at the same level, just as CAC is under the Central Committee, again highlighting the Party nature of the CAC network. Unfortunately, information concerning the structure, funding and missions of local cyber authorities is also incomplete.

The CAC and, where they exist, local CAC websites offer scant insight into the local CACs, other than the top leadership and scattered, and not always up-to-date, public events. The Guangdong Provincial CAC website, for example, discloses the names of the director and two deputy directors, announces some local activities such as training and policy-related meetings, but does not report any enforcement actions. More information is available on some local CACs through publicly-disclosed annual budget reports, which are not available for CAC itself. These describe in varying detail the main functions, institutional structure, staffing, and income and expenditures of those local CACs, but this information is largely a matter of speculation—other than its functions—in the case of the CAC.

While CAC is an active rule maker, issuing rules to regulate on a nationwide basis the sectors and activities subject to its authority, the local CACs are not. However, both CAC and the local CACs conduct enforcement activities, covering a range of actions, with a jurisdictional division of authority. CAC leads on issues of national scope, working with other relevant departments, for example, on cybersecurity reviews. For seemingly most enforcement matters, the CAC network implements the territorial principle, with local CACs taking general and specific instruction or 'guidance' from higher levels with respect to companies and matters falling within their regions. For example, the CAC requested Beijing CAC to summon the online question-and-answer platform Zhihu for publishing illegal information, after which Beijing CAC filed an administrative punishment case against Zhihu (Global Times 2021). CAC reportedly first admonished, and then instructed the Beijing CAC to fine,

China's leading social media platform Sina Weibo US$470,000 for allegedly repeatedly publishing or transmitting illegal information, the forty-fifth and largest fine so imposed against the company in 2021 (Song 2021). That fine surpassed the over US$1.4 billion in fines for content transgressions that the Beijing CAC imposed that year on popular online entertainment discussion platform Douban.com (Lin 2021), in which Beijing CAC, upon orders from CAC, stationed inspectors in March 2022 to deal with "serious online chaos" (CAC 2022c). As another example, CAC outlined a series of more internet content actions planned for 2022 that task local CACs, as well as key website platforms, to formulate work plans based on CAC's rectification priorities to ensure unified standards and actions (CAC 2022a).

CAC and the local CACs also rely on sectoral government departments to be involved in or on the front line for many regulatory actions (CAC et al. 2021). For example, the Shenzhen Municipal CAC in Guangdong province, jointly with the local public security, market supervision and transportation authorities, met with more than twenty Internet companies to discuss and have them sign publicly-disclosed pledges to better protect personal information on their apps (CAC 2021e). CAC also looks to the public to help with enforcement, including through the Reporting Centre for Illegal and Unhealthy Online Information that receives complaints and passes them along to individual online operators, monitoring their handling. Illustratively, in April 2022, the Centre accepted nearly 440,000 reports, the vast majority of which concerned the Weibo platform (CAC 2022d). Local CACs also receive reports from the public, amounting to 806,000 instances in the same month (CAC 2022e)

## The Legality of CAC's Role

CAC's relationship to its regulated public is complicated. Under the above mentioned principle of "principal responsibility," it has assigned companies increasing responsibility for the content on and management of their websites. It often initially takes a "soft" enforcement approach that prioritises compliance over punishment, first summoning one or more companies to admonish them concerning various unlawful behavior and seek commitments to rectify the behaviours, or else face fines and possibly harsher punishment. Such commitments are often publicised to increase social pressure on the companies. However, CAC can also act seemingly arbitrarily, as in the Didi case, where it did not provide a clear legal rationale for its punishment decisions as other ministries tend to do. It released little information about what exactly triggered the reviews or their process, which continued for just over one year, generating much uncertainty for the companies involved and the broader market (Horsley 2022). CAC, through its investment fund with the Ministry of Finance, CIIF, has taken shares in regulated companies and reportedly may take one in Didi (Reuters 2021), ostensibly to have a more direct voice in their management.

The CAC network is seemingly immune from administrative law requirements on transparency that apply to government agencies, including publishing its structure, finances, powers, and responsibilities, and accountability through an appeal process and litigation. A proposed overhaul of 2017 CAC rules on enforcing internet information content requirements would supplement and expand the coverage of CAC's enforcement authority as provided under the CSL, DSL, and PIPL to apply as well to cybersecurity, data security and personal information protection obligations (CAC 2022f). The draft provisions incorporate some but not all procedural requirements in the Administrative Procedure Law that was substantially revised in 2021, which would to a certain extent constrain the CAC's enforcement authority by providing more protections to the parties subject to enforcement. They, however, do not incorporate disclosure requirements or an obligation to provide compensation to injured parties that are imposed on its regulatory counterparts like MIIT and MPS (MPS 2018). Surprisingly, however, unlike CAC itself, many provincial and lower-level CACs publish their annual budgets and accounts; some local CACs also publish annual open government information reports. Such diversity suggests that the CAC network is in practice decentralised in some respects.

The Party under Xi Jinping has promoted law-based governance to put power in a 'cage of regulations.' As part of its concurrent drive to enhance party leadership, however, it created in CAC an opaque, complex, active, and seemingly unaccountable party regulator with tentacles in many sectors and issues.

## QUESTIONS RAISED

A portrait of the CAC as an institution cannot but remain incomplete. Much information surrounding its composition, tasks, and institutional contexts has not been made public. Yet another problem has, thus far, been the relatively limited bandwidth of foreign analysts trawling through the significant amount of information it produces, or that are published about it in Mandarin. Furthermore, CAC must be understood as an evolving department. Not only is it relatively new itself, with all the consequences that entails, the policy areas over which it presides equally move at breakneck pace, pushing CAC toward continuous adaptation. Further inquiry is thus indispensable, and the following questions provide ample grist for the analytic mill.

## About the CAC as an Individual Regulator

CAC is still a young institution and likely feeling its way along, including in its interactions with other party and state institutions. There is still much to learn about CAC, given the shroud of secrecy under which it operates. It does appear to be both generally acting in accordance with procedures required for state institutions when engaging in rulemaking and most enforcement matters, and also in line with its party nature. It fulfils its 'leadership' and coordination role on behalf of the CCIC, a high-level party deliberative body, in policymaking, regulation and enforcement. However, it looks not only to local CACs, which serve a similar function within their localities, but principally to the functional state regulators to organise and carry out enforcement within their competencies and under guidance from the centre.

But what is its working relationship with those regulators? In the social credit field, the central bank, which shares with the NDRC overall responsibility for developing that mechanism, has reportedly managed to keep its credit reporting information—much of which, as in the West, is confidential and only shared with entities authorised by its customers—separate and distinct from the NDRC's social credit database. The CAC, MIIT, and MPS appear to have achieved a workable division of labour. Are there still bureaucratic complexities and occasional divergence of interests at play in the cyberdigital regulatory space, or does CAC's party nature confer greater status that trumps all others' and helps ameliorate or at least resolve any tensions?

The situation becomes even more opaque when turning to the CAC's embedding in the institutions of the Party Centre. How does it receive instructions, through the also secretive CCIC or directly from the Politburo Standing Committee or from General Secretary Xi himself at times? What is its decision making process? And what is its precise relationship with the CPD, given CAC's mandate to regulate and enforce online information content? How much of CAC's staff and work are devoted to propaganda-related matters? How is it organised and staffed internally? Is it growing still, expanding its remit, or has it settled into an increasingly clear and uncontested role? All these are questions that merit further investigation and, hopefully, some eventual sunlight.

## About Its Role in Realising Chinese Tech Policy

CAC was established to help unify cyberspace regulation and achieve a workable balance between national security and economic development with the innovation that leaders recognise is critical to China's success. Is that happening in practice? Does CAC's party status help or hinder its cooperation with others? Alternatively, does CAC have the clout to compel other Party or State bodies to abide by the legislative and regulatory frameworks it oversees? The Shanghai Public Security Bureau data breach of July 2022 (Arcesati and Hmaidi 2022) could provide an insightful test case in this regard.

Further analysis should also be devoted to the sorts of coalitions that CAC builds up in its attempt to design, implement and iterate technology policy and regulations. Little is known, for instance, about the operations of its own in-house think tank, CACS, in comparison to the MIIT-affiliated China Academy for ICT (CAICT), which issues frequent reports and white papers on the state of digital affairs. How, then, does CAC generate policy suggestions, or where does it acquire them from? How does it work together with other Party and State departments, universities, the technical community, and the private sector? How does it respond to social concerns, as it seems to have done in relation to gig workers and delivery drivers (Sheehan and Du 2022)? One useful case study is that of the National Cybersecurity Centre in Wuhan (Cary 2021), and much more information is available in the public domain to be mined.

On the international stage, more work should be done on the structure of China's foreign engagement and the role that the CAC plays. In a sense, CAC more directly represents the top-level of decision making (Xi Jinping and the Politburo Standing Committee) than the Ministry of Foreign Affairs does. Yet very little information is available on lines of communication between CAC and MFA, how specific foreign policy mandates are created, or possible strategies formed.

Such questions are largely descriptive in nature. Yet the theoretical discussions they inform require our factual knowledge of Chinese institutional processes, including the interplay between Party and State dynamics, to increase drastically. Perhaps the major guiding question in contemporary Chinese studies comes down to how the Chinese Communist Party leadership envisions China's future and attempts to adapt to the exigencies of its circumstances. This is not merely a rational intellectual exercise, but is riven with more prosaic and pragmatic considerations, institutional interests, and human foibles. With its central position in digital affairs, CAC will lie at the heart of China's development agenda for the decades to come. Its functioning, or dysfunctionality, will significantly impact China's economic trajectory, its social stability and its standing in the world. We ignore it at our peril.

## NOTES

1. The official English-language name of this body is the Central Cyberspace Affairs Commission. However, this chapter uses the above translation, which is more faithful to the original Mandarin name.

## REFERENCES


Arcesati, Rebecca and Antonia Hmaidi. 2022. Shanghai Police Database Breach Exposes Lax Data Protection. MERICS, 20 July. Available from: https://merics.org/en/short-analysis/shanghai-police-database-breach-exposes-lax-data-protection.

Boyd, Alexander. 2021. The Historical Nihil-List: Cyberspace Administration Targets Top Ten Deviations from Approved History. *China Digital Times*, 16 August 2022. Available from: https://chinadigitaltimes.net/2021/08/the-historical-nihil-list-cyberspace-administration-targets-top-ten-deviations-from-approved-history.

Boyd, Alexander. 2022. Cyberspace Administration of China Lists 500 Paeans to "Positive Energy. *China Digital Times*, 18 January 2022. Available from: https://chinadigitaltimes.net/2022/01/cyberspace-administration-of-china-lists-500-paeans-to-positive-energy.

CAC. Undated. Zhao Zeliang jianli (CV of Zhao Zeliang). Available from: http://www.cac.gov.cn/bgs/ldhd/xzr/jl/A0902010504index_1.htm.

CAC. 2015. Hulianwang yonghu zhanghao mingcheng guanli guiding (Internet User Account Naming Management Measures), issued 4 February. Translation available from: https://chinacopyrightandmedia.wordpress.com/2015/02/04/internet-user-account-name-management-regulations.

CAC. 2017. Wangluo chanpin he fuwu anquan shencha banfa (shixing) (Interim Security Review Measures for Network Products and Services), issued 2 May. Translation available from: https://chinacopyrightandmedia.wordpress.com/2017/05/02/interim-security-review-measures-for-network-products-and-services.

CAC. 2017a. Guojia wangluo anquan shijian yingji yu'an (National Cybersecurity Incident Emergency Response Plan), issued 10 January. Available from: http://wlaq.xjtu.edu.cn/info/1009/1013.htm.

CAC. 2019. Zhongguo Hulianwang Touzi Jijin: ziben zhuli wangxin qiye xingwen zhiyuan (China Internet Investment Fund: Capital Assists Cybersecurity and Informatization Enterprises to Operate Steadily and Grow Far). 30 October. Available from: http://www.cac.gov.cn/2019-10/30/c_1573967258883096.htm.

CAC. 2019a. Ertong geren xinxi wangluo baohu guiding (Measures on the Online Protection of Childrens' Personal Information), issued 22 August. Available from: http://www.cac.gov.cn/2019-08/23/c_1124913903.htm.

CAC. 2021. Guanyu jinyibu yashi wangzhan pingtai xinxi neirong guanli zhuti zeren de yijian (Opinions concerning Further Consolidating the Principal Responsibility of Website Platforms for Information Content Management), issued 15 September. Available from: http://www.gov.cn/zhengce/zhengceku/2021-09/16/content_5637727.htm.

CAC. 2021a. Wangluo shuju anquan guanli tiaoli (zhenqiu yijian gao) (Online Data Security Management Regulations [Draft for Comment]), issued 14 November. Translation available from: https://digichina.stanford.edu/work/translation-online-data-security-management-regulations-draft-for-comment-nov-2021.

CAC. 2021b. Qiche shuju anquan guanli ruogan guiding (shixing) (Some Measures on Vehicle Data Security Management [Trial]), issued 16 August. Available from: http://www.cac.gov.cn/2021-08/20/c_1631049984897667.htm.

CAC. 2021c. Cyberspace Administration of China Launches the Initiative on China-Africa Jointly Building a Community with a Shared Future in Cyberspace, 25 August. http://www.cac.gov.cn/2021-08/25/c_1631480920680924.htm.

CAC. 2021d. Fahui shuzi jishu youshi gongtong tuidong jianpin shiye fazan (Give Rein to the Advantages of Digital Technology, Jointly Promote the Undertaking of Poverty Reduction and Development), 11 June. Available from: http://www.cac.gov.cn/2021-06/11/c_1624994157738454.htm.

CAC. 2021e. Guangdongsheng Shenzhen shiwei wangxinban tuidong zhongdian hulianwang qiye gongkai chengnuo 'baohu geren xinxi'yashi zhuti zeren (Shenzhen Municipal Party Committee of Guangdong Province CAC promotes key Internet companies to make public commitments to 'protect personal information' to consolidate their main responsibility). China Netcom, 15 November. Available from: http://www.cac.gov.cn/2021-11/15/c_1638574277375399.htm.

CAC, 2022. Shuju chujing anquan pinggu banfa (Outbound Data Transfer Security Assessment Measures), issued 7 July. Translation available from: https://digichina.stanford.edu/work/translation-outbound-data-transfer-security-assessment-measures-effective-sept-1-2022.

CAC. 2022a. 2022 nian 'qinglang' zhuanxiang xilie xingdong juxing xinwen fabu hui (Press Conference Held on the 2022 "Qinglang" Series of Special Actions), 18 March. Available from: http://www.cac.gov.cn/2022-03/17/c_1649125522577850.htm.

CAC. 2022b. Zhongyang Wangxin Ban zhaokai quanguo wangluo wenming jianshe gongzuo tuijinhui (CAC Convenes National Online Civilization Construction Work Advancement Conference), 27 June. Available from: http://www.cac.gov.cn/2022-06/27/c_1657946521680409.htm?mc_cid=ce7ba52dc5&mc_eid=4f18c05255.

CAC. 2022c. Wangxin bumen gongzuo dudaozu jinzhu Douban Wang (Cyberspace Authority Supervision Team Stationed at Douban.com), 15 March. Available from: http://www.cac.gov.cn/2022-03/15/c_1648951412843416.htm?mc_cid=980d833c86&mc_eid=4f18c05255.

CAC. 2022d. Zhongyang Wangxinban Jubao Zhongxin fabu zhuyao shangye wangzhan pingtai 2022 nian 4 yuefen wangluo qinquan jubao shouli chuzhi qingkuang (CAC Reporting Centre Publishes Details on Online Infringement Report Acceptance and Processing concerning Major Commercial Website Platforms for April 2022), 11 May 2022. Available from: http://www.cac.gov.cn/2022-05/11/c_1653880789797463.htm.

CAC. 2022e. 2022 nian 4 yue quanguo shouli wangluo weifa he buliang xinxi jubao 1602.9 wan jian (16 Million Reports on Illegal and Harmful Information Online Received Nationwide in April 2022). 18 May 2022. Available from: http://www.cac.gov.cn/2022-05/18/c_1654486479793333.htm.

CAC. 2022f. Wangxin bumen xingzheng zhifa chengxu guiding (zhenqiu yijian gao) (Provisions on Procedures for Administrative Law Enforcement by Cyberspace Departments [Draft for Seeking Comments]), issued 8 September. Available from: http://www.cac.gov.cn/2022-09/08/c_1664174174624227.htm.

CAC, AQSIQ (Administration for Quality Supervision, Inspection and Quarantine) and SSMC (State Standardization Management Commission). 2016. Guanyu

Jiaqiang guojia wangluo anquan biaozhunhua gongzuo de ruogan yijian (Some Opinions concerning Strengthening National Cybersecurity Standardization Work), issued on 12 August. Available from: http://www.cac.gov.cn/2016-08/22/c_1119430337.htm.
CAC et al. 2020. 2020 nian shuzi xiangcun fazhan gongzuo yaodian (2020 Outline for Digital Rural Development Work), issued 9 May. Available from: http://www cac.gov.cn/2020-05/08/c_1590485983517518.htm.
CAC et al. 2021. Wangluo anquan shencha banfa (Cybersecurity Review Measures), issued 28 December. Translation available from: https://digichina.stanford.edu/work/translation-cybersecurity-review-measures-revised-effective-feb-15-2022.
Caiping, Liu and Denise Jia. 2021. Analysis: Didi's Fate and China's Overseas Share Sale Policies. Caixin, 6 July. Available from: https://www.caixinglobal.com/2021-07-06/analysis-didis-fate-and-chinas-overseas-share-sale-policies-101736325.html.
Cary, Dakota. 2021. China's National Cybersecurity Center: A Base for Military-Civil Fusion in the Cyber Domain, CSET, July. Available from: https://cset.georgetown.edu/publication/chinas-national-cybersecurity-center.
CCIC. 2021. "Shisiwu" guojia xinxihua guihua ("14th Five-Year Cycle" Plan for National Informatization), issued on 28 December. Available from: https://www.gov.cn/xinwen/2021-12/28/5664873/files/1760823a103e4d75ac681564fe481af4.pdf.
Central Committee 2017. Zhongguo Gongchandang gongzuo jiguan tiaoli (shixing) (Party Central Committee issues Regulations on Work Organs of the Chinese Communist Party [Trial]), issued 12 April. Available at: http://www.xinhuanet.com/politics/2017-04/12/c_1120797991.htm.
Central Committee. 2018. Shenhua Dang he guojia jigou gaige fang'an (Plan for Deepening the Reform of Party and State Institutions), issued 1 March. Available from: http://www.gov.cn/zhengce/2018-03/21/content_5276191.htm#1.
Central Committee. 2020. Zhongguo Gongchandang Zhongyang Weiyuanhui gongzuo tiaoli (Work Regulations of the Central Committee of the Chinese Communist Party), issued 12 October. Available from: http://www.xinhuanet.com/politics/2020-10/12/c_1126597105.htm.
Chen, Stella and David Bandurski. 2022. CNKI's Security Problem. China Media Project, 6 July. Available from: https://chinamediaproject.org/2022/07/06/cnkis-security-problem.
CIOC (Central Institutional Organization Commission). 2014. Zhongyang bianban zhengcefaguisi jigou bianzhi gongzuo yongyu (shiyong) (Department of Policy and Regulations of the CIOC Explanation of Work Terms in Institutional Organisation [Trial]), issued November. Available from: http://www.hljorg.gov.cn/pages/Article.aspx?ID=42917.
CIOC. 2015. Guanyu Gongye he Xinxihua Bu youguan zhize he jigou tiaozheng de tongzhi (Notice Concerning Adjustment of the Ministry of Industry and Information Technology's Relevant Duties and Bodies), issued 20 April. Translation available from: https://digichina.stanford.edu/work/notice-concerning-adjustment-of-the-ministry-of-industry-and-information-technologys-relevant-duties-and-bodies.

China Copyright and Media 2018. Xi Jinping's Speech at the National Cybersecurity and Informatization Work Conference. Digichina, 22 April. Available from: https://chinacopyrightandmedia.wordpress.com/2018/04/22/xi-jinpings-speech-at-the-national-cybersecurity-and-informatization-work-conference.

CIUHIRC (China Internet Unlawful and Harmful Information Reporting Centre). 2014. Zhongguo Hulianwang Weifa he Buliang Xinxi Jubao Zhongxin de jianjie (Brief Introduction of the China Internet Unlawful and Harmful Information Reporting Centre). 2 September. Available from: https://web.archive.org/web/20180801190640/http://www.12377.cn/txt/2014-09/02/content_7198763.htm.

Creemers, Rogier. 2017. Cyber China: Upgrading Propaganda, Public Opinion Work and Social Management for the Twenty-first Century. *Journal of Contemporary China* 26(103), 85–100.

Creemers, Rogier. 2022. "Cybersecurity Law and Regulation in China: Securing the Smart State." *China Law and Society Review*. OnlineFirst.

Creemers, Rogier. 2022a. China's Emerging Data Protection Framework. *Journal of Cybersecurity* 8(1).

CSAC. 2019. Zhongguo Wangluo Kongjian Anquan Xiehui zhangcheng (Charter of the Cyberspace Security Association of China), issued on 2 December. Available from: https://www.cybersac.cn/News/getNewsDetail/id/87/type/41.

Daum, Jeremy. 2022. Children and the Law in China: An Overview of Recent Reforms (Working Paper Draft). China Law Translate. Available from: https://www.chinalawtranslate.com/en/working-paper-children-and-the-law-in-china-an-overview-of-recent-reforms.

DigiChina. 2022. Knowledge Base: China's 'Global Data Security Initiative' 全球数据安全倡议, 31 March. Available from: https://digichina.stanford.edu/work/knowledge-base-chinas-global-data-security-initiative.

DigiChina. 2022a. Translation: Chinese Authorities Announce $1.2B Fine in DiDi Case, Describe 'Despicable' Data Abuses. DigiChina, 21 July. Available from: https://digichina.stanford.edu/work/translation-chinese-authorities-announce-2b-fine-in-didi-case-describe-despicable-data-abuses.

Economist. 2021. China's Communist Authorities are Tightening Their Grip on the Private Sector. 18 November. Available from: https://www.economist.com/business/chinas-communist-authorities-reinvent-state-capitalism/21806311.

Fang, Jianwei and Wenpei Yu. 2020. China Apps Governance in 2019: Retrospect and Suggestions. Zhong Lun, 21 January. Available from: http://www.zhonglun.com/Content/2020/01-21/1425120471.html.

Gao, Charlotte. 2019. 'Double-Faced' Lu Wei Jailed for 14 Years for Bribery. *The Diplomat*. 27 March. Available from: https://thediplomat.com/2019/03/double-faced-lu-wei-jailed-for-14-years-for-bribery.

Global Times. 2021. China's Cyberspace Regulator Summons Quora-like Platform Zhihu for Publishing Illegal Information. *Global Times*. 20 December. Available from: https://www.globaltimes.cn/page/202112/1242950.shtml.

Guancha. 2014. Zhongyang Wangluo Anquan he Xinxihua Lingdao Xiaozu chengyuan mingdan, 12 zhengfu guoji jianzhi shengaizu (Name List of the Members of the Central Leading Group for Cybersecurity and Informatization, 12 Full and Vice

National-Levels Also Appointed to Leading Group for Deepening Reform), 28 February. Available from: http://www.guancha.cn/politics/2014_02_28_209672.shtml.

Guangming Daily (2014). Zhongguo Hulian Wangluo Xinxi Zhongxin (CNNIC) renshi tiaozheng, Li Xiaodong danren zhuren (Personnel Adjustment at the China Internet Network Information Centre (CNNIC), Li Xiaodong Appointed Chairman). 29 December. Available from: https://news.sina.cn/cm/2014-12-29/detail-iavxeafr9478753.d.html.

Horsley, Jamie. 2022. Behind the Facade of China's Cyber Super-Regulator: What We Think We Know—and What We Don't—about the Cyberspace Administration of China. DigiChina, 8 August. Available from: https://digichina.stanford.edu/work/behind-the-facade-of-chinas-cyber-super-regulator.

Kan, Michael. 2014. Facebook Welcomes Chinese Regulator to US, Even as Site Remains Blocked. *Computer World*. 8 December. Available from: https://www.computerworld.com/article/2856435/facebook-welcomes-chinese-regulator-to-us-even-as-site-remains-blocked.html.

Lin, Hongchao. 2019. Dangzheng jiguo ronghe yu xingzheng fa de huiying (The Integration of Party and Government Institutions and the Response of Administrative Law). CUPL, 10 September. Available from: http://fzzfyjy.cupl.edu.cn/info/1035/11063.htm.

Lin, Liza. 2021. China Fines Weibo for Spreading 'Illegal Information.' *Wall Street Journal*, 14 December. Available from: https://www.wsj.com/articles/china-fines-weibo-for-spreading-illegal-information-11639482120.

Marczak, Bill et al. 2015. China's Great Cannon. Citizen Lab, 10 April. Available from: https://citizenlab.ca/2015/04/chinas-great-cannon.

MFA. 2020. Global Initiative on Data Security. 8 September. Available from: https://www.mfa.gov.cn/ce/ceus/eng/zgyw/t1812951.htm.

Mozur, Paul. 2015. China's Internet Censorship Anthem Is Revealed, Then Deleted. *New York Times*, 12 February. Available from: https://archive.nytimes.com/sinosphere.blogs.nytimes.com/2015/02/12/chinas-internet-censorship-anthem-is-revealed-then-deleted.

MPS. 2018. Gong'an jiguan banli guojia peichang anjian chengxu guiding (Provisions on the Procedures for Handling State Compensation Cases by Public Security Organs), issued 31 December. Available from: http://www.gov.cn/zhengce/zhengceku/2018-12/31/content_5428642.htm.

NPC. 2016. Zhonghua Renmin Gongheguo wangluo anquan fa (Cybersecurity Law of the People's Republic of China), issued 7 November 2016. Translation available from: https://digichina.stanford.edu/work/translation-cybersecurity-law-of-the-peoples-republic-of-china-effective-june-1-2017.

NPC. 2021. Zhonghua Renmin Gongheguo shuju anquan fa (Data Security Law of the People's Republic of China), issued 10 June. Translation available from: https://digichina.stanford.edu/work/translation-data-security-law-of-the-peoples-republic-of-china.

NPC. 2021a. Zhonghua Renmin Gongheguo geren xinxi baohu fa (Personal Information Protection Law of the People's Republic of China), issued 20

August. Translation available from: https://digichina.stanford.edu/work/translation-personal-information-protection-law-of-the-peoples-republic-of-china-effective-nov-1-2021.

Reuters. 2021. Fretting about Data Security, China's Government Expands Its Use of 'Golden Shares,' 16 December. Available from: https://www.reuters.com/markets/deals/exclusive-fretting-about-data-security-chinas-government-expands-its-use-golden-2021-12-15.

Sheehan, Matt and Du, Sharon. 2022. How Food Delivery Workers Shaped Chinese Algorithm Regulations. Carnegie Endowment for International Peace, 2 November. Available from: https://carnegieendowment.org/2022/11/02/how-food-delivery-workers-shaped-chinese-algorithm-regulations-pub-88310.

Shu, Catherine. 2014. China Tried to Get World Internet Conference Attendees to Ratify This Ridiculous Draft Declaration. *TechCrunch*, 20 November. Available from: https://techcrunch.com/2014/11/20/worldinternetconference-declaration.

Song, Ren. 2021. Beipi weifa weigui fabu chuanshu xinxi, Xinlang Weibo zai aifa 300 wan Yuan (Sina Weibo Fined 3 Million Yuan Again for Publishing or Distributing Information Violating Laws and Regulations). VOA Asia, 15 December. Available from: https://www.voachinese.com/a/Weibo-fined-by-Chinese-regulator-for-publishing-illegal-information-20211214/6354452.html.

State Council. 2014. Guanyu shouquan Guojia Hulianwang Bangongshi fuze hulianwang xinxi neirong guanli gonzuo de tongzhi (Notice concerning Empowering the Cyberspace Administration of China to Be Responsible for Internet Information Content Management Work), issued 26 August. Translation available from: https://chinacopyrightandmedia.wordpress.com/2014/08/26/notice-concerning-empowering-the-cyberspace-administration-of-china-to-be-responsible-for-internet-information-content-management-work.

State Council. 2019. Guanyu cujin pingtai jingji guifan jiankang fazhan de zhidao yijian (Guiding Opinions concerning Stimulating the Sound and Healthy Development of the Platform Economy), issued 8 August. Available from: http://www.gov.cn/zhengce/content/2019-08/08/content_5419761.htm.

Webster, Graham and Triolo, Paul. 2020. Translation: China Proposes 'Global Data Security Initiative.' New America, 7 September. Available from: https://www.newamerica.org/cybersecurity-initiative/digichina/blog/translation-chinese-proposes-global-data-security-initiative.

Wei, Lu. 2013. Concentrate Positive Energy Online, Build the Chinese Dream Together (Speech at the 13th Chinese Online Media Forum), 30 October. Translation available from: https://chinacopyrightandmedia.wordpress.com/2013/10/30/lu-wei-concentrate-positive-energy-online-build-the-chinese-dream-together.

Lu, Wei. 2014. Gongxiang de wangluo, gongzhi de kongjian (A Network Shared Together, a Space Governed Together), 23 June. Translation available from: https://chinacopyrightandmedia.wordpress.com/2014/06/23/a-network-shared-together-a-space-governed-together.

WIC. Undated. Hulianwang zhi guang bolanhui (Light of the Internet Expo). Available from: https://expro.wicwuzhen.cn/#/about.

WIC. 2015. Wuzhen changyi (Wuzhen Initiative), issued 18 December. Translation available from: https://digichina.stanford.edu/work/wuzhen-initiative.

WIC. 2017. "Yidai yilu" shizi jingji guoji hezuo changyi (Proposal for International Cooperation on the "One Belt, One Road" Digital Economy), issued 3 December. Translation available from: https://chinacopyrightandmedia.wordpress.com/2017/12/03/proposal-for-international-cooperation-on-the-one-belt-one-road-digital-economy.

WIC. 2022. Xi Sends Congratulatory Letter to Inauguration of World Internet Conference Organization, 13 July. Available from: https://www.wuzhenwic.org/2022-07/13/c_788406.htm.

Wright, Teresa. 2017. Political Discourse on the Internet in China: A Multifarious Virtual Space. In Hansson, Eva and Wright, Meredith, eds., *Political Participation in Asia: Defining and Deploying Political Space*. Abingdon: Routledge, 133–150.

Xi, Jinping. 2015. Speech at the 2nd World Internet Conference Opening Ceremony, 16 December. Translation available from: https://chinacopyrightandmedia.wordpress.com/2015/12/16/speech-at-the-2nd-world-internet-conference-opening-ceremony.

Xinhua. 2011. Guojia hulianwang xinxi bangongshi jiu bangongshi sheli ji qi zhize dawen (State Internet Information Office Answers Questions on the Office's Establishment and Its Duties). 5 May. Available from: http://www.gov.cn/jrzg/2011-05/05/content_1858131.htm.

Xinhua, 2015. Spotlight: China Believes in UN's Leading Role in Coordinating States' Cyberspace Behavior, 17 April. Available from: http://www.china.org.cn/world/Off_the_Wire/2015-04/17/content_35344330.htm.

Xinhua 2020. Zhongyang Wanxinban: Lianhe xiangguan bumen daying wangluo fupin shouguan zhan (CAC: Winning the Online Poverty Relief Battle Together with Relevant Departments), 13 November 2020. Available from: http://www.xinhuanet.com/politics/2020-11/13/c_1126734026.htm.

Xinhua, 2022. Xi Sends Congratulatory Letter to Inauguration of World Internet Conference Organization. 13 July. Available from: https://english.www.gov.cn/news/topnews/202207/13/content_WS62ce206bc6d02e533532da86.html.

Xinhua, 2022a. China Focus: State Council Committee Stresses Economic, Financial Stability. Xinhua, 16 March. Available from: https://english.news.cn/20220316/3bf0096ed0ad4bd98f84b77ca4c7f22a/c.html.

Xinhua, 2022b. Cong wangluo daguo xiang wangluo qiangguo kuobu mijin—"Zhongguo zhe shinian" xilie zhuti xinwen fabuhui ju xinshidai wangluo qiangguo jianshe chengjiu (From a Big Internet Country to a Strong Internet Country—A Series of Press Conferences on the Theme of "This Decade of China" on the Achievements of Building a Strong Internet Country in the New Era), 20 August. Available from: http://www.gov.cn/xinwen/2022-08/20/content_5706135.htm.

# Chapter 2

# The Stumbling Smart State

## *Fragmented Policy Experimentation and Dubious Consolidation*

Straton Papagianneas and Adam Knight

### INTRODUCTION

China, like all modern states, has sought to introduce digital and data-driven practices into its domestic governance as it adapts to the challenges of the twenty-first century. Evolving out of an emerging scientism in the 1980s, this process of informatisation (*xinxihua*) can be understood as one element of a broader package of neoliberal tools and techniques applied in public management (Pieke 2009; Gewirtz 2022; Bray and Jeffreys 2016). Governance in China's post-Tiananmen era has been characterised by its application of more complex, diffuse, and immersive methods—both foreign and home-grown—to the heart of China's state building project, modernising, and strengthening the Party's leading role in society in the process (Shue and Thornton 2017; Bray and Jeffreys 2016). A major goal of this process has been the automation of certain aspects of public administration as a way of stimulating economic activity and streamlining bureaucracy, all without compromising on the Party's absolute authority on matters of ideology and morality. As with the rollout of other key government initiatives, the construction of China's 'Smart State' has relied on a process of experimentation whereby overall principles are set centrally, but the specifics of execution are trialled at the local level (Knight 2020). While providing significant benefit in terms of adaptability, policy outcomes are often hampered by 'implementation gaps,' as entrenched technical, legal, and political standards and interests frustrate the smooth realisation of central objectives (Chen and Greitens 2022).

This chapter assesses the progress of China's 'Smart State' through the lens of such fragmented policy experimentation by examining two case studies, the Social Credit System (SCS) and Smart Court Reform (SCR). Both examples—in particular the SCS—have received relatively wide coverage in English-language media, though little work has been done to situate the systems within broader patterns of governing practice and informatisation.

Anglophone coverage of the SCS has grown in recent years, owed in part to significant—though often problematic—media coverage of the topic. A significant strand of this literature is rooted in the framework of 'authoritarian resilience' and questions of whether big-data enabled projects such as social credit will strengthen or weaken the Chinese Communist Party's (CCP) control over society (Chen and Cheung 2017; Hoffman 2018; Liang et al. 2018). A limited number of empirical studies have been carried out to date, mostly based on survey data to show levels of public awareness and acceptance of social credit (Kostka 2019). Similarly, a handful of case study–based projects have illuminated specific applications of the system on-the-ground (Knight 2020; Knight and Creemers 2021). Some limited work has been done to situate the SCS within China's broader governance and reform agenda, upon which this chapter will build (Creemers 2018; Zhang 2020).

English-language scholarship on SCR has tended to focus on its legality and functional purposes (Peng and Xiang 2020; Wang and Tian 2022a). Other scholars have depicted SCR as a means to strengthen central control over judicial power (Zheng 2020; Stern et al. 2021). In general, English-language scholarship has expressed its concerns for the potentially undermining effect of automation and digitisation to judicial adjudication (Shi et al. 2021). In contrast, Papagianneas (2021b) found a generally positive attitude toward SCR and their effect on justice in the Chinese language scholarship. This is explained by the positivist organisational and ideological principles of Marxism-Leninism: technology and automation provide a way forward toward achieving the dream of rational Marxist governance (Munro 1971; Bakken 2000; Hoffman 2017). This is why automation projects such as the SCS and SCR are so enthusiastically embraced by both central and local state actors.

This chapter asks a simple question: Where are we now? In this short overview, we discuss and assess the latest regulatory developments of the SCS and SCR. In addition, we examine example case studies as a way to discuss the Smart State 'in action,' concluding that they ought to be viewed as part of a broader attempt to recentralise vertical governing power through technology. The first section will explore the origins and principles of both systems, paying particular attention to their roots in the desire to automate elements of public administration that gathered pace from the 1990s through to the 2010s. The second section will then examine the processes through which the SCS and SCR have been implemented as part of a broader pattern of governance-through-experimentation, as well as the unique challenges posed by technology. Finally, the third section will document domestic critiques of both the SCS and SCR, as well as reforms carried out since 2020 to consolidate and strengthen the 'Smart State' in these areas.

## THE CSC AND THE CSR: PRINCIPLES AND ORIGINS

The desire for standardisation, informatisation, and automation is deeply connected with the emergence of the modern nation-state and bureaucracy (Porter 1995). Both in public and corporate governance, numerical indicators have become the primary tool to achieve efficiency and accountability (Demortain 2019). The CCP has an extra affinity with quantification and automation due to its interpretation of Marxist-Leninist ideology. This ideology holds that social reality is reducible to a set of objective truths that simply exist and are waiting to be extracted (Munro 1971; Hua 1995; Bakken 2000). It underscores the importance of a vanguard institution, finding these objective truths and transforming them into actionable decisions to lead the masses on a path of national progress. The vanguard institution uses this input-driven decision making process to control the masses and simultaneously adapt its capacity to maintain this control, with the ultimate goal to sustain itself as a so-called benevolent and efficient rule. Therefore, the CCP blends public participation with top-down control, which allows it to constantly shape, manage, and respond to society and itself (Hoffman 2017; Gueorguiev 2021). The SCS and SCR are the latest iteration of the digitisation and automation of this governance process.

### The Social Credit System

The definition and direction of the SCS has evolved significantly throughout its existence. Receiving its first high-level political mention in 2002, the search for a credit system with Chinese characteristics began at least a decade earlier, as economists and policymakers alike sought to resolve early existential threats to China's nascent market economy. The opportunity of Reform and Opening brought with it significant risk. Private business was encouraged, but China's regulatory system was ill-equipped to mediate such new commercial relationships. The potential cost of defaulting on contractual promises was so low that cases of fraud reached near-endemic levels (Yan 2021; Lee 2014). Banks sought out new tools and indicators to determine the risk profile of borrowers without a history of financial transactions and behaviour (see Chorzempa's chapter in this book). The goal of the original architects of the SCS was simple: to find a solution that mitigated these risks and institutionalised the kind of commercial trust required for the free flow of finance, goods, and services in marketized economies.

To achieve this, policy researchers from the Chinese Academy of Social Sciences turned to credit systems around the world for inspiration, in particular the United States. Yet while the SCS may have had its roots in Western financial practice, the system quickly evolved to take on several uniquely Chinese features. These developments mapped neatly onto other prevailing trends in Chinese governance and politics as a response to a wide range of regulatory challenges, most notably the revival of traditionalist strands of virtue-based rule and the elevation of the Socialist Core Values across all aspects of law and governance (Gow 2017; Creemers and Trevaskes 2020). In the early 2010s, the definition of 'creditworthiness' or 'trustworthiness' (*chengxin*) was expanded beyond the merely financial to incorporate additional meaning in the social, judicial, and governmental realms. This was codified in the State Council's 2014 *Planning Outline for the Construction of a Social Credit System*, often referred to as the first true SCS document (Creemers 2018). The *Planning Outline* described a credit system whose scope covered not only economic but also social management, encouraging and discouraging a wide range of behaviours and sectors, from taxation to transportation, and the environment to education.

At its core, the SCS can be distilled to a single principle multiplied across the many jurisdictions in which it is operational. The system's guiding logic is to ensure that 'those deemed untrustworthy in one area shall be restricted everywhere' (*yichu shixin, chuchu shouxian*). Participating ministries at both the central and local level maintain 'blacklists' (*hei mingdan*) of entities deemed to have violated relevant rules within the jurisdiction of that particular authority. Details of blacklisted entities are then published online through the department's own website, as well as on the national level 'Credit China' platform managed by the NDRC. Businesses and individuals can search these databases for offending parties, while departments undertake to mutually recognise and jointly impose 'disciplinary measures' (*chengjie*) within their own jurisdiction (known as the 'joint punishment system' (*lianhe chengjie zhidu*) through a network of MoUs. The goal here is twofold: (1) to increase the cost of 'untrustworthy' (*shixin*) behaviour through additional layers of punishment so as to (2) gradually transition from a postevent regulatory regime to a preprevention model in which 'untrustworthiness' is reduced across the board (Shen Y. 2019).

In social credit, we see an attempt to use information to increase accountability for one's actions—both directly through punishments, as well as indirectly through reputational damage—and for the fear of such accountability to encourage compliance with rules and directives. In delegating such control down to the individual person or entity, we should understand the SCS as part of a broader push to streamline and even automate governance in China. Indeed, the *Planning Outline* explicitly states that a core goal of the system is to urgently 'reduce administrative governmental interference' in the economy and society. Subsequent high-level social credit documents have repeatedly linked the construction of the system with the delegation of control as part of a wider transformation of governing techniques. This includes an emphasis on social credit as a tool for greater enforcement of judicial decisions, as well as 'social governance,' a practice that differs from previous attempts at 'social management' through its emphasis on 'co-construction, -governance, and -sharing' with a variety of actors, both public and private, as well as an increasing reliance on principles of individual self-governance (Snape 2019; Ma 2018).

## Smart Court Reform

Smart Court Reform (SCR) started in 2016 when Chief Justice Zhou Qiang delivered the Annual Working Report of the Supreme People's Court (SPC). He mentioned that the smart court system should

> Make full use of technologies such as the internet, cloud computing, big data, artificial intelligence and so on, to promote the modernization of trial system and judgement capability, so as to achieve the highly intellectualised operation and management of people's court.

Subsequently, the SPC published the *Five-Year Development Plan for the Informatisation of People's Courts* ('Development Plan'), the 2016 *Opinion on Comprehensively Promoting the Synchronous Generation and In-depth Application of Electronic Archives*, and the 2017 *Opinion on Accelerating the Building of Smart Court* ('2017 SPC Opinion'). The main task of informatisation at the time was primarily to support other judicial reforms, such as the circuit-courts, improving judicial services, increasing judicial responsibility, and expanding and improving channels of oversight in courts. SCR is only one part of a series of unprecedented vast and broad reforms of the entire judiciary, which started in 2013.

At a basic level, the goal of SCR is to create courts where judicial officers use technological applications to facilitate their internal and operational judicial and administrative work, provide better judicial services to the public, and improve enforcement of judicial decisions. The term 'smart court' (zhihui fayuan) is used to indicate any (physical or online) court where the judicial process is conducted on a digital platform. This platform is integrated with advanced applications based on algorithms, AI, and big data analytics, which allows for the automation of specific judicial processes. They are enthusiastically embraced by both frontline and senior judges because they facilitate the day-to-day work of the former and the oversight and management tasks of the latter (Stern et al. 2021; Papagianneas 2022).

The first step of SCR was to digitise the entire judicial process (i.e., case-submission, trial preparation, trial hearing, issuing of judgment, serving court documents). It improved and facilitated the work of judicial officers (e.g., frontline judges, senior supervisors, and court leadership). Full digitisation allows people to submit cases via the internet or via automated dockets in court halls, therefore improving access to justice (Xu 2017). It also allows a complete recording of every procedural step and the real-time monitoring of frontline judges' work by their superiors. This possibility improved trial management and oversight by the court leadership over their subordinates, which also improved uniform adjudication (Papagianneas 2022).

The second step came when full digitisation of the judicial process allowed for the automation of certain processes. The possibility of AI independently adjudicating complex (politically or socially) sensitive cases remains minimal, as this is the discretion of the Chinese Communist Party (CCP). Nonetheless, AI is used in other ways: software application exist that can automatically index the facts of a case, match it with similar legal cases, provide applicable legislation and regulations, and give recommendations to the case-handling judge on how to rule, based on big-data analysis of similar cases (Faxin 2020; Ma 2020). Another example is the use of AI to adjudicate similar cases (e.g., online financial borrowing and small loan contract disputes) automatically in bulk. These kinds of applications are integrated in courts' digital case management platform (Guo 2019).

Nonetheless, automation refers not so much to the automation of adjudication. Rather, in SCR discourse, automation refers to the reduction of human agency in the making of discretionary decisions during the judicial process: the *Development Plan* implies that the end-goal of informatisation, and, therefore, SCR is to build a 'systemic iron cage' or a 'digital big-data iron cage' around adjudicators. The 2017 *Opinion* states that smart courts should promote 'the organic unification of substantive and procedural justice.' This implies that digitisation should improve the adherence to procedures, but that these procedures remain in service of substantive outcomes. Together with other judicial reforms, SCR is about improving and better enforcing judicial procedures at the cost of human discretion. Judicial reformers believe that this makes the judiciary more efficient, more consistent, and, therefore, fairer (Hu 2019).

According to a research report in 2022, the third phase of People's Court Informatisation has been officially completed. It means that smart courts can conduct all judicial operations completely online, have achieved full disclosure of the judicial process through digitisation, and are able to provide all-round intelligent services. By the end of 2021, electronic or online litigation was used in eighteen percent of judicial trials nationwide, which is a seventeen percent-point increase from 2016. The next phase, People's Court Informatisation 4.0 will be about building 'all-round intelligence, full system integration, full business collaboration, full ubiquity over space and time, and full system autonomy' (Wang and Tian 2022b)

In the next section, we give an overview of the way that SCS and SCR have been implemented and the consequent issues.

## IMPLEMENTATION: FRAGMENTED EXPERIMENTATION

Governance in China is no monolith. The size and diversity of the country's geography and population present unique challenges, while its political system lacks some of the inherent mechanisms for popular feedback and reform (such as elections or a free media) that are found in democratic systems. As a result, the Chinese government has needed to find alternative methods to build in the kind of agility and responsiveness necessary to adjust to an increasingly complex world of issues. A model of 'adaptive governance' has emerged, wherein guerrilla-style policy experiments that blend central visions with local realities allow the state to deal creatively with pervasive uncertainty (Heilmann and Perry 2011).

Both the SCS and SCR have relied heavily on a highly decentralised model of decision making and localised piloting as part of their rollout (Knight 2020). Foundational documents such as the 2014 SCS *Planning Outline* and the 2017 SPC *Opinion on Smart Courts* are purposefully light on detail beyond providing a general direction of travel, with decisions as to how those priorities should be executed devolved across all levels of government as a way of fostering policy innovation, appraising cadre performance, and shielding higher authorities from potential public criticism.

In the case of the SCS, this has led to a system that is best described as a network of networks, consisting of many hundreds of interconnected systems built between 2014 and 2020 that while underpinned by the same guiding logic, operate largely independent of each other (Liu 2019). At the central level, at least forty-seven bodies are currently involved in the system's design and management, collectively publishing thousands of individual documents (Drinhausen and Brussee 2021). At the top of this pyramid sits the NDRC and PBoC, co-leads of the InterMinisterial Joint Conference on the Construction of the Social Credit System, as well as the State Council in charge of coordinating cross-departmental collaboration. Beneath this are more than fifty

MoUs guaranteeing mutual recognition of blacklists and joint punishment had been released (Wu and Liu 2020). In addition to centrally administered systems, hundreds of local schemes have emerged since 2014, each vying for acknowledgement by their superiors at the provincial and national level. Three sets of model pilot cities have been published since 2016 and most recently in September 2021, with case studies of successful implementation circulated and then emulated across the country. Each of these systems maintains its own blacklists and redlists. A small minority have incorporated some degree of point scoring in their municipal systems. Within certain parameters, these actors have historically determined what behaviours should and should not be included, how data should be collected and stored, and what punishments or rewards should be applied. These lists are then (sometimes, but not always) shared with other localities through a series of MoUs and provincial-level data-sharing agreements and technical interfaces. This fragmentation of the system has led to a bewildering array of social credit applications in response to specific, often localised governance challenges.

SCR has followed a similar trajectory. Courts in China are responsible for adhering to reform objectives on their own. Given the great disparity between courts in terms of finances (Ng and He 2017), some have bigger budgets for costly digitisation projects than others. Courts across the country have started digitising their operations way before the SCR was formally launched in 2015–2016. It is only from then on that this transformed into a top-down driven policy, starting with the designation of two pilot programs in the Jilin and Zhejiang High Courts (Xu 2017).

The provincial high courts took the lead in developing smart systems that were then implemented in intermediate and basic people's courts. In addition, courts developed their new systems in cooperation with a private partner, such as Alibaba or iFlytek. One of the first courts to develop an artificial intelligent system for the judicial process, was the Shanghai High Court. It developed a 'trial centred litigation reform software' in partnership with iFlytek. The system was originally meant for criminal cases but has now expanded to civil and commercial cases as well, and is used across courts in Shanghai (Cui 2020). In Zhejiang, the Hangzhou Internet Court, developed an online e-commerce court platform in cooperation with Alibaba (Mingay 2019). While it shows the government's willingness to work with private industry to implement reform, it also exacerbates the issues of fragmented policy implementation, as we will see below.

## The Stumbling Smart State: Emerging Critiques

This model of implementation has been central to the rollout and innovation of both the SCS and SCR, providing maximum agility and responsiveness while also shielding the central government from criticism should the systems have met with public pushback. This honeycomb-like pattern of siloed schemes with differing technical standards and practices has, however, caused no end of problems when it comes to integration at the regional or national level. Such unfettered expansion has come at the expense of uniformity and moderation, causing bottlenecks in the systems' standardisation that threaten their continued rollout, as well as their legitimacy in the eyes of policymakers and the wider public.

At the heart of this issue lies the critique that programmes such as the SCS and SCR have become overly 'generalised' (*fanhua*), incorporating all manner of technological efforts outside of their original scope. This has led to accusations of policy short-cuts, with 'lazy' officials 'hijacking' the SCS and SCR as vehicles for their short-term goals to avoid the more arduous process of creating actual legislation (Wang 2020). In the case of the SCS, this has led to numerous examples of system overreach, wherein localities introduce new behaviours or incentives into the SCS without any legal grounding, essentially introducing a system of extrajudicial punishment and reward. This has been particularly controversial in the twenty or so municipalities where points-scoring mechanisms have been constructed. Many of the system's earliest architects have looked on with a degree of horror as the SCS has expanded in this way since 2014, fearing that such a lack of legal foundation risks undermining the overall legitimacy of the SCS (Knight 2022).

Similarly, in the case of SCR, the term 'smart courts' has provided rhetorical cover for all manner of technological applications, from the most basic digitisation efforts (e.g., enabling digital filing) to the automation of processes with sophisticated algorithmic software (e.g., automatic analysis of cases and pushing of relevant legislation, past decisions, and sentence recommendation to adjudicating judges). Just as with the SCS, the introduction of such techniques outpaced their incorporation into law. Without a coherent legal framework, there exists no strong legal basis for the digitised judicial process, as procedural laws do not yet recognise the legal validity of electronic versions of submitted evidence, witness statements, etc. While local courts, such as the Internet Courts had issued relevant documents for digital processes, such as e-filing, they did not have national effect. Therefore, concerns have emerged that this legal uncertainty and inconsistent regional regulations could undermine the credibility and ambition of the smart courts (Peng and Xiang 2020).

The legal issues created by these uses of technology in governance have, in the minds of many scholars and officials, undermined access to justice and fairness in China. Of course, official discourse claims that the SCS and SCR will only increase judicial accountability and fairness. These kinds of statements typify a kind of technological solutionism common among Chinese officials. In the case of SCR, however, the reality is that its emphasis on a more efficient, standardised, consistent, and politically controlled justice system risks reducing the importance of legal interpretation, judicial discretion, and consideration of individual circumstances. In this sense, technology will have a dehumanising effect on justice administration, potentially affecting perceptions of fairness among litigants. It also has the potential to undermine the agency of judges, turning them into mere law-applying bureaucrats, rather than law-interpreting professionals (Ji 2018; Sun 2019). Therefore, automation has serious implications for judicial pluralism. Especially in a unitary judicial system such as China's, technology does not have to fully replace human judges to have a dehumanising effect or significantly reduce human discretionary decision making.

Such legal issues aside, another key problem caused by the rapid growth and 'generalisation' of Social Credit and the Smart Courts has been a lack of technical interoperability between systems. This issue came to a head during China's response to the COVID-19 pandemic in particular (Knight and Creemers 2021). On the one hand, the localised nature of the SCS allowed for a rapid retooling of some of the system's features to adapt to the unfolding public health crisis. Certain features of the SCS were paused temporarily, while new behaviours such as facemask-wearing and isolation were folded into the system's enforcement mechanisms. Yet this flexibility also proved a weakness as the transregional nature of the pandemic required increased levels of data-sharing. Local administrations quickly found that social credit systems built to differing technical standards as part of their rapid rollout were unable to 'talk' to each other, sometimes even within the same province. This was not a new problem; indeed, ever since the launch of the SCS, the breakup of 'information islands' (*xinxi gudao*) has been a key reform priority. The COVID-19 crisis amplified these voices and catalysed a process of centralisation and reform to which we will return in the next section of this chapter.

This push for greater technical integration of the sharing of data between different nodes of China's Smart State has naturally led to significant privacy and cybersecurity challenges, however. During the first wave of the pandemic, analysis by Chinese state media found that only three out of fourteen provincial 'health code' systems included any kind of provisions for the protection of personal data (The Paper 2020). Similar issues have plagued SCR from the very outset. For example, in 2013, the SPC launched a public database for court decisions as a build-up to SCR (Ahl and Sprick 2017). The digitisation of all court decisions was an important first step to provide the big data for machine learning. Later, smart systems were connected to these kinds of databases to support automated suggestions to judges. Courts were tasked to upload as many of their decisions onto it. However, initially, courts did not remove the personal information of litigating parties, including minors, in criminal cases, divorce, and custody cases (Liebman et al. 2019). This was only done later as new national rules require smart courts to align their data-management practices with new personal information protection laws.

## CONSOLIDATION STAGE

Considering the above issues and the potentially existential threat they pose to the continued development of the Smart State, the last two years have seen a concerted effort by many scholars and policymakers to bring reform to both the SCS and SCR. Indeed, after a period of rapid expansion and experimentation, both systems have now firmly entered a phase of consolidation and reform, with the primary focus of introducing greater regulatory or procedural standardisation at the national level.

In the case of the SCS, the last two years have seen a raft of new regulations published with the goal of upgrading the system beyond the original 2014 *Planning Outline*. This shift has come against a backdrop of an increasingly hawkish stance among the SCS's key planning bodies. In August 2019, the deputy director of the NDRC Policy Research Office stated, 'We have noticed that [the SCS in] some places violates laws and regulations by incorporating behaviours that are not applicable within the scope of the punishment mechanism for untrustworthiness within personal credit records. We are correcting and dealing with the situation without further delay' (Credit China 2019). The spokesperson went on to lay out a strategy of 'three prevents,' namely to avoid the generalisation and expansion (1) of what is defined as an untrustworthy behaviour and their incorporation into credit records, (2) of further blacklists and other punishment measures, and (3) of the creation of further credit-building measures such as personal credit points and scores (Credit China 2019). The goal was to create a SCS that sits within China's legal system, not in parallel to it. This rhetorical shift was quickly matched in terms of legislative updates, with five major new documents published by the State Council, NDRC, and PBoC between July 2020 and March 2022 (Knight 2022). Since 2019, progress has also been made toward the creation of a Social Credit Law, with multiple symposia between policymakers, academics, and industry held to discuss its design.

Taken in sum, these updates have sought to rein in the SCS at its fringes, curtailing the excesses of its phase of fragmented experimentation. They clarify what data should be collected and classified within the remit of the SCS, when those data should be shared publicly and how, what punishments could be applied, and how one's credit record could be appealed and altered. The new draft rules look to further standardise blacklisting and punishments in particular, ensuring that any disciplinary measures taken are rooted in law and are not overly punitive. If officials feel that a particular law is not adequately tough, they must lobby for changes to that law rather than simply fabricating their own administrative punishments through the SCS. Crucially, these documents stipulated that all local or ministerial systems would be evaluated in due course, with noncomplying versions then shut down.

Likewise, SCR has seen the introduction of regulatory and procedural standardisation at the national level. In 2021 and 2022, the SPC introduced national rules to standardise and unify SCR. In quick succession, the SPC issued the *Online Litigation Rules* (OLR), *the Online Mediation Rules* (OMR), *the Online Courts Operation Rules* (OCOR), and the *Opinions on Strengthening the Judicial Application of Blockchain* (Blockchain Opinion). The publication of these documents indicate that the stage of consolidating experiences and unifying practice has begun. They are most likely the first step in standardising smart court procedures. In the future, we might see the development of a national law related to online procedures, on par with the Civil and Criminal Procedure Laws (Papagianneas 2021a). These documents aim to unify and standardise the smart systems as well as their application, operation, and management (OCOR, article 1). They ask for more coordination and planning from the top (OCOR, article 2.3), which is a strong signal of more centralised planning and coordination. The Blockchain Opinion also signals a focus on improving interconnectivity, collaboration, and information-sharing between courts and other sectors and standardising the use of blockchain systems in the judicial system at a national level (Deng 2022).

These updates have sought to address key concerns around human agency and control in particular. New national regulations aim to provide increased agency to litigants during the digital judicial process, giving them a concrete sense of control over the process. Litigants have, for example, the right to choose the method of litigation (online or offline) (OLR, Article 2). It obliges courts to obtain the explicit consent of litigating parties and inform them of their rights and obligations, the practicalities, and legal consequences of online litigation (OLR, Article 4). It allows parties to separate procedures between online and offline, that is, consenting to conducting part of the judicial process online may not be seen as consent to conducting the entire judicial process online (OLR, Article 4.4). Consent to online litigation may be revoked at any time during the judicial process, and the court is obliged to transfer the process back to offline if it does not find any objections (OLR, Article 5). The courts may also conduct the process on a double offline—online track if one of the parties does not consent to online litigation (OLR, Article 10). The OLR also has multiple provisions that allow litigants to maintain control over and access to the judicial process as much as possible: Article 14 and 20 allow parties to participate in the litigation process at separate times within a certain time period. Therefore, litigants' procedural rights are protected, at least on paper.

Despite the criticisms regarding human agency and discretionary decision making of judges, SCR is clearly intended to increase consistency and efficiency through enhanced vertical control (Zheng 2020; Stern et al. 2021). Therefore, the consolidation stage has not seen a reversing or addressing of the issue, rather a continuation of this chosen path. For example, smart systems at both local and provincial-level, in combination with local and national regulations, have standardised and institutionalised the well-known 'trial management and oversight' mechanism. This mechanism allowed senior court leadership to intervene in politically and socially sensitive cases but was frequently abused for personal gain and a significant source of judicial corruption (He 2012; Li 2012; Wang 2020). Through the standardisation, digitisation, and then automation of specific processes of the 'trial management and oversight' mechanism, these smart systems enabled a stricter and more consistent application of the mechanism, where the system also monitored every step undertaken by both supervised judge and supervising court leader. Automation of justice, in a sense, does not necessarily refer to the replacement of humans in the judicial process, but rather a significant reduction of human agency in the judicial decision making process.

## CONCLUSION

The conception, construction, and consolidation of the social credit and smart court systems has been emblematic of China's Smart State rollout more broadly. In these systems, we see a reflection of the CCP's evolving governing logic, fusing views of informatisation as an answer to governance and legal challenges with a reassertion of state-arbitered morality and the Party's 'leadership over everything.' Since the mid-2010s, two of the government's flagship 'smart' projects—the SCS and SCR—have relied heavily on a decentralised model of experimentation at the departmental and local level as a way of fostering innovation at speed. This has led many local administrations to interpret and apply the meaning and methods of social credit and smart courts as they saw fit, in response to a range of governance challenges. This 'generalisation' of the Smart State led to a mission creep that proved highly controversial, with many questioning the system's legality and thereby legitimacy. An increasing clamour for reform has led to renewed legislative attention, as the government upgrades and bolsters both social credit and smart courts in order to ensure their longevity.

After a period of relative decentralisation in terms of the systems' rollout and experimentation, the process of recentralisation should be understood as part of a broader effort to leverage artificial intelligence and automation to increase vertical control of governance and society as a core goal of the Smart State. In the case of SCR in particular, while China's judiciary believes that taking a leading (global) role in AI will give them leverage against unwanted incursion from the party-state, it in fact does the opposite (Hall et al. 2005). It has become clear that SCR is a way for the central judiciary to increase vertical control over their courts and judges. Various initiatives are explicitly oriented towards reducing the discretionary decision making power of human judges by embedding their work process in a tightly circumscribed and digitally surveilled environment. Therefore, SCR is meant to reshape the judicial bureaucracy into a legally rational institution by institutionalising channels of political control.

As the advancement of the Smart State rolls on to the point in which defining between 'smart' and 'non-smart' elements of government becomes moot, the issues highlighted in this chapter—the legal and practical impact of tech-enabled governance—will only become more prescient. These are global challenges, but the speed with which China is looking to informationalise its governing practice requires an expedited response.

## REFERENCES

Ahl, B. and Sprick, D. 2017. Towards judicial transparency in China: The new public access database for court decisions. *China Information* 32, 3–22.

Bakken, B. 2000. *The Exemplary Society: Human Improvement, Social Control and the Dangers of Modernity in China*. Oxford: Oxford University Press.

Bray, D. and Jeffreys, E. 2016. *New Mentalities of Government in China*. Routledge.

Chen, Huirong, and Greitens, S. 2022. Information capacity and social order: The local politics of information integration in China. *Governance* 35, 497–523.

Chen, Yongxi, and Cheung, A. 2017. The Transparent Self Under Big Data Profiling: Privacy and Chinese Legislation on the Social Credit System. *The Journal of Comparative Law* 12, 356–78.

Credit China. 2019. 'NDRC: Prevent the Generalization of Untrustworthy 'Blacklists" [国家发改委□要防止失信'黑名单'认定的泛化]. Available at https://www.creditchina.gov.cn/xinyongfuwu/shixinheimingdan/heimingdanyanxi/201908/t20190816_165743.html.

Creemers, R. 2018. China's Social Credit System: An Evolving Practice of Control. http://dx.doi.org/10.2139/ssrn.3175792.

Creemers, R. and Trevaskes, S. 2020. *Law and the Party in China: Ideology and Organisation* Cambridge: Cambridge University Press.

Cui, Yadong. 2020. *Artificial Intelligence and Judicial Modernization*. New York: Springer Publishing.

Demortain, D. 2019. The Politics of Calculation: Towards a Sociology of Quantification in Governance. *Revue d'Anthropologie des Connaissances* 13, 973–90.

Deng, Beixian. 2022. 'SPC Releases New Policy on Blockchain Technology in Judicial Field.' *China Justice Observer*. Accessed 14 September 2022. https://www.chinajusticeobserver.com/a/spc-releases-new-policy-on-blockchain-technology-in-judicial-field.

Drinhausen, K. and Brussee, V. 2021. China's Social Credit System in 2021: From fragmentation towards integration. *Mercator Institute for China Studies*. https://merics.org/en/report/chinas-social-credit-system-2021-fragmentation-towards-integration.

Faxin. 2020. '法信2.0智推系统 (Faxin 2.0 Smart Push System).' Accessed 6 Dec 2020. http://www.faxin.cn/html/about/fxzt/Fxzt.aspx.

Gewirtz, J. 2022. *Never Turn Back: China and the Forbidden History of the 1980s*. Boston: Harvard University Press.

Gow, M. 2017. The Core Socialist Values of the Chinese Dream: Towards a Chinese integral state. *Critical Asian Studies* 49, 92–116. 10.1080/14672715.2016.1263803.

Gueorguiev, D. 2021. *Retrofitting Leninism: Participation Without Democracy in China*. Oxford: Oxford University Press.

Guo, Wenli. 2019. The Four Major Judicial Innovations of China's Guangzhou Internet Court. *China Law Connect* 6, 1–6.

Hall, M, Calabro, D, Sourdin, T, Stranieri, A. and Zeleznikow, J. 2005. Supporting discretionary decision making with information technology: Case study in the criminal sentencing jurisdiction. *University of Ottawa Law Technology Journal* 2, 1–36.

He, Xin. 2012. Black Hole of Responsibility: The Adjudication Committee's Role in a Chinese Court. *Law & Society Review* 46, 681–712.

Heilmann, S. and Perry, E. 2011. *Mao's Invisible Hand: The Political Foundations of Adaptive Governance in China*. 1 ed. Cambridge: Harvard University Asia Center.

Hoffman, S. 2017. 'Programming China: The Communist Party's Autonomic Approach to Managing State Security.' Ph.D. dissertation, The University of Nottingham.

Hoffman, S. 2018. Social Credit: Technology-enhanced authoritarian control with global consequences. *Australian Strategic Policy Institute*. https://apo.org.au/node/180186.

Hu, Yue. 2019. 'Cui Yadong: Rengong zhineng rang sifa gengjia gongzheng (Cui Yadong: Artificial Intelligence Makes Justice Fairer).' *Fenghuang wang*, 30 August 2019. Accessed 25 May 2022. https://perma.cc/7NZ4-7ZCX.

Hua, Shiping. 1995. *Scientism and Humanism: Two Cultures in Post-Mao China* (1978–1989). Albany, NY: State University of New York Press.

Ji, Weidong 2018. 人工智能时代的司法权之变 [Changes in Judicial Power in the Era of Artificial Intelligence]. 东方法学 [*Eastern Legal Studies*] 1, 125–33.

Knight, A. 2020. 'Technologies of Risk and Discipline in China's Social Credit System.' In *Law and The Party in China: Ideology and Organisation*, edited by Creemers, R. and Trevaskes, S. 237–62. Cambridge: Cambridge University Press.

Knight, A. 2022. 'Basket Case: Socialist Law and the Reform of China's Social Credit System.' *China Law and Society Review*. Forthcoming.

Knight, A, and Creemers, R. 2021. Going viral: The social credit system and COVID-19. *Available at SSRN 3770208*.

Kostka, G. 2019. China's social credit systems and public opinion: Explaining high levels of approval. *New Media & Society* 21, 1565–93.

Lee, Haiyan. 2014. *The Stranger and the Chinese Moral Imagination*. Stanford University Press.

Li, Ling. 2012. The 'production' of corruption in China's courts: Judicial politics and decision making in a one-party-state. *Law & Social Inquiry* 37, 848–77.

Liang, Fan, Das, V, Kostyuk, N. and Hussain, M. 2018. Constructing a data-driven society: China's social credit system as a state surveillance infrastructure. *Policy & Internet* 10, 415– 53.

Liebman, B, Roberts, M, Stern, R. and Wang, A. 2019. Mass Digitization of Chinese Court Decisions: How to Use Text as Data in the Field of Chinese Law. *Journal of Law and Courts* 8, 177–201. http://dx.doi.org/10.2139/ssrn.2985861.

Liu, Chuncheng. 2019. Multiple Social Credit Systems in China. *Economic Sociology: The European Electronic Newsletter* 21, 22–32.

Ma, Cong. 2020. "Faxin 2.0 zhitui xitong' shangxian! Quanmian duijie quanguo fayuan dianzi juanzong ban'an xitong ['The Faxin 2.0 Smart Push System' is online! Fully connected with the national court electronic case handling system].' *Smart Court Times*, 30 October 2020. Accessed 30 March 2022. https://www.anpcn.com/hangyezixun/566.html.

Ma, Qingyu. 2018. 'An Interpretation of the Connotations of the Coconstruction, Co-governance, Co-sharing Social Governance Setup' [共建共治共享社会治理格局的意涵解读]. Administrative Management Reform [行政管理改革].

Mingay, A. 2019. 'Size matters: Alibaba shapes China's first 'Court of the Internet."' *Mercator Institute for China Studies*. Accessed 17 October 2019. https://merics.org/en/analysis/size-matters-alibaba-shapes-chinas-first-court-internet.

Munro, D. 1971. The Malleability of Man in Chinese Marxism. *The China Quarterly* 48, 609–40.

Ng, Kwai Hang, and Xin He. 2017. *Embedded Courts: Judicial Decision-Making in China*. Cambridge: Cambridge University Press.

Papagianneas, S. 2021a. 'Brief: Consolidating Digital Justice.' *China Law Translate*. Accessed 02 March 2021. https://www.chinalawtranslate.com/en/brief-consolidating-digital-justice.

Papagianneas, S. 2021b. Towards Smarter and Fairer Justice? A Review of the Chinese Scholarship on Building Smart Courts and Automating Justice. *Journal of Current Chinese Affairs* 51, 327–47. https://doi.org/10.1177/18681026211021412.

Papagianneas, S. 2022. Automating Intervention in Chinese Justice: Smart Courts and Supervision Reform. *Asian Journal of Law and Society* Forthcoming.

Peng, Junlin, and Wen Xiang. 2020. The Rise of Smart Courts in China. *NAVEIÑ REET: Nordic Journal of Law and Social Research* 1, 345–72.

Pieke, F. 2009. *The good communist: Elite training and state building in today's China:* Cambridge University Press.

Shen, Yilong. 2019. 'On Administrative Joint Disciplinary Measures for the Untrustworthy and its Legal Control' [论失信的行政联合惩戒及其法律控制]. *The Jurist [法学家]* 4.

Shi, Changqing, Sourdin, T. and Bin Li. 2021. The Smart Court—A New Pathway to Justice in China? *International Journal for Court Administration* 12, 1–19.

Shue, V, and Thornton, P. 2017. *To govern China: Evolving practices of power*: Cambridge University Press.

Snape, H. 2019. Social Management or Social Governance: A Review of Party and Government Discourse and why it Matters in Understanding Chinese Politics. *Journal of Chinese Political Science* 24, 685–99. https://doi.org/10.1007/s11366-019-09605-2.

Stern, R, Liebman, B, Roberts, M, and Wang, A. 2021. Automating Fairness? Artificial Intelligence in the Chinese Courts. *Columbia Journal of Transnational Law* 59, 515–53.

Sun, Daocui 2019. 我国刑事司法智能化的知识解构与应对逻辑 [Knowledge Deconstruction and Corresponding Logic of China's Criminal Judicial Intelligence]. 当代法学 *[Contemporary Legal Studies]* 3, 15–26.

The Paper 2020. '14省市健康码仅3地有知情同意和隐私保护条款' ['Out of 14 Provincial and Municipal Health Codes, Only Three Have Informed Consent and Privacy Protection Clauses'], April 30, 2020, accessed December 5, 2020, https://www.thepaper.cn/newsDetail_forward_7210904.

Wang, Lu 2020. 'Discussing Generalized Credit Over Ten Days' [泛化信用十日谈]. *Caixin Online [财新网]*. Available here https://opinion.caixin.com/2020-06-10/101565405.html.

Wang, Nyu, and Tian, M. 2022a. "Intelligent Justice': AI Implementations in China's Legal Systems.' In *Artificial Intelligence and Its Discontents. Social and Cultural Studies of Robots and AI*, edited by Hanemaayer, A. 197–222. London: Palgrave Macmillan.

Wang, Yiming, and He Tian. 2022b. 'The development of China's court informatization in 2021 and the outlook for 2022.' In *China Court Informatization Development Report No. 6*, edited by Guoping Chen, He Tian, Yanbin Lu and Changming Hu. Beijing: Social Sciences Literature Press.

Wang, Yueduan 2020. Overcoming Embeddedness: How China's Judicial Accountability Reforms Makes Its Judges More Autonomous. *Fordham International Law Journal* 43, 737–66.

Wu, Yulin and Heng Liu 2020. 'Memoranda of Understanding on Credit Joint Punishments and Rewards: Operational Logic, Legal Nature and Approach to Rule of Law' [信用联合奖惩合作备忘录：运作逻辑、法律性质与法治化进路]. *Social Science Journal of Henan* [河南社会科学] 28(3), 11–20.

Xu, A. 2017. Chinese judicial justice on the cloud: a future call or a Pandora's box? An analysis of the 'intelligent court system'of China. *Information & Communications Technology Law* 26, 59–71.

Yan, Jun 2021. Social Credit System Based on Personal Benefit Records of Social Insurance. *Financial Engineering and Risk Management* 4, 91–95.

Zhang, Chenchen 2020. Governing (through) trustworthiness: technologies of power and subjectification in China's social credit system. *Critical Asian Studies* 52, 565–88.

Zheng, G. 2020. China's Grand Design of People's Smart Courts. *Asian Journal of Law and Society* 7, 561–82.

# PART II

# Strategic Emerging Technologies

# *Chapter 3*

# China's Industrial Policy for Semiconductors

John Lee

Semiconductors are foundational to modern electronics. As the basis for integrated circuits (ICs), they shape the possibilities offered by almost every category of contemporary and emerging technology. This brings implications for national power and security that run against the transnational character of the semiconductor value chain. The central role in this picture of China, and of China's increasingly antagonistic relations with the United States and its allies, has made semiconductors a focus of the new 'geopolitics of technology.' Understanding the Chinese state's approach to semiconductors, and China's aggregate position within the global semiconductor value chain, is required to assess the overall international balance of power and the prospects for China's Party-State to achieve its long-term goals (MERICS 2021).

A dominant position in globalised high technology industries provides a source of structural power in the international system (Malkin 2022; McCarthy 2015). ICs were invented in the United States, and their production process has retained strong continuities over the decades, only with rising levels of complexity and specialisation. Accordingly, the influence of US firms and the presence of US-owned technology throughout the semiconductor industry remains high and provides an instrument through which the US state can exercise power against China. As described by senior US officials (White House 2022; Office of the USTR 2022), the US export controls of October 2022 targeting China's semiconductor industry are meant to 'freeze' or at least severely constrain China's progress, thereby sustaining US technological leadership and the preponderance of power that this embeds.

Yet this goal seems to not be accepted uncritically by US-allied states whose firms hold key positions in the semiconductor value chain (Bloomberg

2022; SCMP 2022). The structure of this industry limits Chinese prospects for dominance but also provides Beijing with sources of power that Washington will find difficult to constrain. As a unique example of technological catch-up by a developing country in the era of 'asymmetrical globalisation,' and in view of semiconductors' crucial role in the Chinese Party-State's' goals, China's place in this industry sheds light on the structural nature of power in the international system and the trajectory of China's national development.

This chapter explains China's place in the globalised semiconductor industry, and its policy goals, development ecosystem, and prospects for advancement in this most foundational of technologies. The discussion highlights major obstacles to Chinese progress that are presented by this industry's features, and the limited capacity to meet these challenges that still characterises China's domestic semiconductor ecosystem. But it also shows how even limited Chinese success in this industry has potentially far-reaching consequences for international politics, even if these falls short of the Party-State's' ambitions by leaving China as a 'partial power' (Shambaugh 2013) in high technology.

The chapter first examines China's involvement in the transnational value chains that typify information technology (IT) industries under 'asymmetrical globalisation,' and in the semiconductor value chain specifically. Second, it reviews the Party-State's' strategic goals that guide its policy for developing the nation's semiconductor industry. Third, it looks at the value chain's objective features, and how these shape priorities in targeting specific processes and technologies. Fourth, it describes China's semiconductor ecosystem—the 'government-research-industrial complex' that drives outcomes in this industry—and China's aggregate position in the global semiconductor value chain. The chapter's conclusion returns to the implications for international power relations, in the context of weaponised interdependence and drift toward strategies of technological containment.

## BREAKING THE MOULD OF 'ASYMMETRICAL GLOBALISATION'

China modernised its IT industries by joining a new model of globalised trade that emerged during the 1990s. This form of globalisation is based on transnational production networks ('global value chains,' GVCs), with advanced economies keeping ownership of core IP and outsourcing lower value-added functions to developing economies (Ernst and Kim 2002). This trend goes far to explain the durability of US international power, which is embedded in the dominance of US firms 'upstream' in GVCs, controlling the core technology that downstream firms require to perform their functions and thereby reaping most of the profits from this economic activity (Malkin 2022).

Since the 1990s, developing economies' participation in GVCs has delivered disappointing results in moving up the technological ladder and so capturing a larger share of value generated by economic activity, resulting in a failure to achieve broad-based wage growth (UNCTAD 2018). This trend has been reinforced by evolution of the WTO trading regime, which has opened developing countries' markets to industry-leading firms from advanced economies, while restricting national autonomy to assist domestic firms with interventionist and protectionist policies. Such measures had previously been used by Japan, South Korea, and Taiwan to help domestic firms to accumulate market share and IP, allowing them to upgrade their technology and thereby their competitive position in globalised industries, notably semiconductors (Matthews and Cho 2009).

China is the salient case of a developing country that, with uneven success, has broken out of this asymmetrical relationship with advanced economies in GVCs. During the 1990s, as China's integration with the global economy proceeded, policymakers tried to push technological upgrading through both command economy style interventions—for example, in semiconductor fabrication (Fuller 2016)—and joint ventures with foreign industry leaders in GVCs. By the early 2000s, these approaches were widely recognised as ineffective (Zhou, Lazonick and Sun 2016, 44). The Party-State adapted its method of involvement in China's economic and technological development to one that has been described as 'grand steerage,' channelling resources through indirect, market-conforming instruments to 'steer' the economy towards broadly defined goals (Naughton 2022).

This approach was still premised on China's integration into a global economy driven by market principles, and into GVCs dominated by foreign firms. But it recognised that 'domestic determinants,' notably state-led industrial policy and the capabilities accrued by domestic firms, are crucial factors in capturing benefits from international trade, including technological advancement (Coe et al. 2004; Ernst 2016; Poon 2018). Chinese state interventionism was enabled by the scale of China's workforce and markets, which attracted foreign industry leaders to China and gave Chinese authorities leverage to impose policies that foreign actors would not accept in smaller developing countries (UNCTAD 2018; Ernst 2016).

These state interventions in the market—creating infrastructure, compensating for firm-level externalities, developing human capital, creating markets for domestic firms, facilitating technology transfer from foreign actors, and subsidising strategic but uncompetitive industries—have created the conditions for China's comparative advantages to be used effectively (Lin and Zhang 2019; Zhou, Lazonick, and Sun 2016). China's exports have shown increasing technological specialisation and rising global market share despite rising domestic factor costs, indicating progressive upgrading by Chinese firms into more sophisticated activities (Malkin 2022: 11). This record contrasts starkly with that of India, which has benefited much less than China from integration with GVCs (Ernst 2016), including specifically in the semiconductor sector (Fuller 2012). India is now adopting a policy approach to semiconductors that looks increasingly like China's (*Economic Times* 2022).

That said, this 'loosely coupled' system of state-led industrial policy has produced varied results across sectors, reflecting variations in market conditions and policy execution (Rho and Kim 2022). In the semiconductor industry, state interventions have frequently been ineffective and wasteful (Fuller 2019). One reason for this is the structure of the transnational semiconductor value chain, which is an extreme case of the asymmetrical concentration within GVCs of market power, and hence of incumbent firms' capacity to maintain control over core IP and high value-add functions.

In 2020, firms headquartered in advanced economies (including Taiwan) captured 95 percent of revenues from the global semiconductor sector, with the United States alone reaping close to 50 percent. This reflects the dominance of US firms in chip design, which accounts for around 50 percent of total value added in the sector. Chinese firms' market share was around 5 percent (SIA 2021). This asymmetry is being amplified by the semiconductor sector's strong growth, estimated at 26.2 percent in 2021 and 16.3 percent in 2022 (WSTS 2022), with robust long-term secular growth predicted (see e.g. Deloitte 2022).

Japan, Taiwan, and South Korea established their firms in the semiconductor value chain through 'industry creation' rather than technology creation: they identified trajectories in technological development and hence markets, then targeted these markets by absorbing existing technology and diffusing it to domestic firms that received extensive state support (Hwang and Choung 2014; Matthews and Cho 2009, 313–314). All three achieved this while the semiconductor value chain was evolving, and so had established their positions by the time the value chain matured in the 1990s.

By contrast, China's industrial policy in this sector delivered lacklustre results prior to the last decade, leaving Chinese firms facing more technically advanced foreign incumbents entrenched across the value chain, many steps in which have high barriers to entry (**Figure 1**). These barriers have grown larger over time with specialisation and rising capital costs, as the demand for ever more computing power has driven semiconductor manufacturing to push the limits of physics (Lee and Kleinhans 2020). One well-known example is photolithography, where performance at the technological frontier is monopolised by a single firm, ASML. ASML's most advanced machines sell for US$150 billion and are a critical tool for cost effective production of recent generation computer processors, due to the precision required to achieve the necessary transistor density.

The negative implications of Chinese industry's weakness in semiconductors have grown in tandem with the technological upgrading of China's export-oriented manufacturing sector, and with the Party-State's' ambitions for China's digital transformation. China's emergence as the global hub for electronics manufacturing has made it the world's largest consumer of semiconductors (McKinsey 2014), resulting in China now spending more on importing semiconductors than oil: in 2020, around US$300 billion (Brookings 2021). Despite China being among the leading locations for IC fabrication, domestic production—two-thirds of which was controlled by foreign (including Taiwanese) firms—accounted for less than 16 percent of China's consumption in 2020 (IC Insights 2021).

Under these conditions, China adopted a 'fast follower' strategy for the semiconductor sector (Verwey 2019), focused on creating conditions for Chinese firms to gain footholds in the different value chain steps and progressively accumulate the capital (financial, human, and technical) to upgrade and become internationally competitive. This approach involved Chinese firms basing their business operations on exchanges with foreign industry leaders, most notably cutting-edge fabrication providers like Taiwan's TSMC. It was hoped that such participation in the semiconductor GVC on asymmetrical terms would produce results comparable to those from Apple locating its manufacturing supply chain in China, which stimulated growth of a competitive Chinese supplier ecosystem and upskilling of China's workforce, notwithstanding the success of Apple and its non-Chinese suppliers in avoiding direct technology transfers to Chinese firms (Grimes and Sun 2016).

Compared with past efforts to catapult selected state-owned firms to the industry's technological frontiers, this gradualist, broad-based, and market-oriented approach is more aligned with that which has arguably defined China's successful cases of technological upgrading (Lin and Zhang 2019). But because semiconductors are such a foundational technology, the Party-State is not satisfied with modest results in this sector over the long term: the ultimate goal is to break free of asymmetrical interdependence with the United States and its allies in the semiconductor value chain, creating the technological basis for China to interact with these states on equal or dominant terms.

## THE PARTY-STATE'S' STRATEGIC GOALS AND THE SEMICONDUCTOR INDUSTRY

### Building the foundations for 'cyberspace superpower'

Over the late 1990s and early 2000s, China's top leaders identified 'informatisation'—the comprehensive application of digital IT—as an organising principle for the nation's development (Austin 2014; Naughton 2002). Since Xi Jinping's accession to the highest leadership positions in late 2012, this has been rearticulated as a vision for making China a 'cyberspace superpower,' with security in cyberspace now given coequal importance with developmental goals (Xinhua 2014). This vision is broadly defined as typical under 'grand steerage,' but it implies that China should develop IT capabilities comparable to the dominant actor in cyberspace, the US (Lee 2022).

Semiconductors are critical enablers for the various systems that constitute cyberspace. Dependence on foreign countries for such 'core technologies' was already identified by Xi Jinping in 2016 as China's 'greatest hidden danger' (Xinhua 2016). This concern was vindicated by the damage done to Huawei, perhaps China's most successful digital technology firm, by US export controls targeting its dependence on foreign semiconductor manufacturing services and software. These and other US measures targeting individual Chinese firms have highlighted the larger Chinese economy's vulnerability to foreign pressure, even in sectors where it has achieved significant progress, due to incapacity to produce foundational components like semiconductors.

As a physical product, semiconductors also belong to the manufacturing-based 'real economy' that has been increasingly emphasised in official rhetoric as the true foundation of national power, and which must be integrated with further development of China's digital economy (Xi Jinping 2021). The Chinese government's much reported 'tech sector crackdown' has focused on internet services firms (MacroPolo 2021), leaving hardware producers generally unscathed.

The National Informatisation strategy released in December 2021 sets out a comprehensive development vision based on integrating the real and digital economies (Oxford Analytica 2022). This document puts ICs at the front of the list of 'core technologies' for which major breakthroughs in addressing shortcomings and building innovation capacity should be made by 2025 (Cyberspace Administration of China 2021). It represents the next stage in Chinese industrial policy's turn toward securing the 'commanding heights' of next-generation technologies, which if achieved would flip the asymmetrical character of GVCs to China's advantage.

## Supporting Leadership in Next-Generation Technologies

By the early 2000s, lacklustre results in sectors such as semiconductor fabrication (Fuller 2016, 122–125) led Chinese policymakers to recognise the poor prospects for technology transfer under conditions of asymmetrical globalisation. This stimulated the policy drive for 'indigenous innovation' to bootstrap domestic technological progress, and the promulgation in 2006 of a fifteen-year 'National and Medium Long-Term Plan (NMLTP) for Science and Technology' (To 2022, 74–75; Zhou, Lazonick, and Sun 2016), which included list of sectoral 'mega-projects' with one dedicated to ICs (Lee and Kleinhans 2021a, 12). However, the NMLTP still recognised multiple innovation pathways and the benefits of incorporating foreign technology (Cheung 2018, 309–311).

Policy evolved again under Xi's leadership from the early 2010s, responding to the need for upgrading China's development model in the face of accumulating economic and demographic pressures, and to the new political imperative to show a 'great rejuvenation of the Chinese nation' through tangible metrics such as technological progress (To 2022; MERICS 2021). In 2015 the 'Made in China 2025' (MiC-25) plan set out ambitious industrial upgrading and import substitution goals for multiple sectors focusing on emerging technologies, and the accompanying industry roadmap addressed various semiconductor-related technologies. The focus on building high technology industries also justified perpetuation of an investment and supply-side driven approach to economic growth, instead of rebalancing the economy towards consumption, which would have required more radical and politically risky changes to China's political economy (Naughton 2022).

Growing pressure on China's access to critical technology inputs from abroad has reinforced the urgency of upgrading domestic industry. China's current (2021–2025) Five-Year Plan lists semiconductors as one of seven 'frontier technologies' prioritised for breakthroughs. In September 2022, a top-level statement was issued on the need for a whole-of-society, Party-led mobilisation to make breakthroughs in 'key core technologies,' albeit still within a market framework (Xinhua 2022). This was followed by the writing of S&T into the Party's constitution at the 20th Congress in October 2022, cementing its place in the 'mission statement' justifying the Party's rule over China.

## ICT Supply Chain Security

Huawei's targeting over 2019–2020 by US export controls highlighted how China's prowess in digital technology remains fundamentally insecure, due to foreign firms' dominance in the key 'chokepoints' of the semiconductor value chain (Figure 1). Cutting-edge fabrication is a duopoly of Taiwan's TSMC's and South Korea's Samsung. Most IC design globally is still based on foreign-owned intellectual property, especially instruction set architectures (ISAs) owned by US (Intel, AMD) or UK/Japanese (Arm) companies. The chip design process uses specialised software (Electronic Design Automation, EDA) tools for which three US-based companies dominate the market, with Chinese EDA vendors accounting for barely 10 percent of China's EDA market in 2020 (Lee and Kleinhans 2021a, 25–29). US, Japanese, and European firms dominate production of semiconductor manufacturing equipment (SME).

These chokepoints were targeted by extensive US export controls issued in October 2022, restricting business with China in categories of advanced ICs, cutting-edge fabrication, and SME (Department of Commerce 2022). Such business is now subject to US government–issued licences with a general presumption of denial, although some exception is made for operations in China by firms headquartered in US-friendly countries. These new controls are unprecedented in applying to China as a jurisdiction rather than to individual firms, and in their extension to 'US persons.' This latter measure deters US firms from continuing business with Chinese customers through offshoring or shell companies and targets the important role in China's semiconductor industry of individuals with US citizenship or residency, who are effectively being forced to choose countries.

While US policymakers are not yet aiming to force a complete 'decoupling' with China, their statements indicate that the new controls fit within a strategy of unqualified technological containment. Quoting the US National Security Adviser (White House 2022), the goal is now to maintain as a large a US technological lead over China as possible, as a national security imperative. This implies active measures to hinder China's technological development, without consideration for economic consequences (CNAS 2022). Washington's express expectation is that allied states will bring their policies and laws in line with this goal, with discussions in progress as of November 2022 with European and East Asian governments whose firms occupy important roles in the value chain.

The viability of a 'fast follower' approach based on unrestricted access to inputs, investment, and partnerships from abroad is now in question, given the proven effectiveness of US assertion of extraterritorial jurisdiction in forcing foreign industry leaders like TSMC to stop business with Chinese firms. But the continued limitations of Chinese industry leave the Party-State with no choice but to continue promoting integration into global technological innovation systems (China-cer.com.cn 2021), hoping to reconcile this by 'pulling tight' international supply chains into dependence on China (Xi

Jinping 2020). Chinese firms are thus trying to maintain or expand foreign business relations, even if some activities like acquisitions are increasingly unfeasible, while progressively introducing domestic suppliers and investing in promising start-ups for this purpose.

MiC-25 set a goal of 40 percent self-sufficiency in IC production by 2020, but by one recent estimate, China will have reached only half this figure by 2025 (IC Insights 2021). China's import substitution rate as of 2020 for SME has been estimated at under 30 percent for all but one among ten SME categories, with the rate for five categories judged to be under 10 percent (Great Wall Glory Securities 2022). For EDA tools, the import substitution rate is generally estimated as remaining under 10 percent. By 2030, these dependencies are likely to be reduced but far from eliminated (Lee and Kleinhans 2021a).

One reason for this slow progress with import substitution is the exponential rise in the Chinese economy's demand for semiconductors. This has encouraged much investment in China's semiconductor sector to be directed at meeting immediate requirements, rather than at long-term technological progress. For example, from 2017 to 2022, SME worth US$93 billion was shipped to China, more than to any other region over the same timeframe. Most if not all this equipment was for use in trailing-edge fabrication, which also reflects existing US export controls on cutting-edge SME and stockpiling by Chinese firms to risk mitigate against future expansion of export controls.

## SPECIFIC GOALS: TRACKING PRIORITIES FOR STATE INTERVENTION IN THE SEMICONDUCTOR INDUSTRY

The general goals described above are *subjective* priorities of China's Party-State. Decisions by Chinese authorities about more specific development priorities are likely to be shaped by *objective* national interests, which are influenced by the characteristics of distinct steps in the semiconductor value chain. Lee and Kleinhans (2021a, 7–11) identify eight such steps, which can be mapped against objective national interests, as shown in Figure 3.1 (dividing the fabrication step into 'cutting-edge' and 'trailing-edge,' given their differing characteristics). Redder colours represent a higher degree of objective importance to the national interest. This in turn indicates the likelihood of state intervention in the market, to promote desired outcomes from a partisan national viewpoint.

Figure 3.1 must therefore be interpreted in the context of states' differing situations and interests. For example, the Assembly, Test, and Packaging step (bottom row) has high importance in terms of espionage risk. The concentration of global ATP activity in China makes this of low concern to Chinese authorities, but of high concern to other states wary of Chinese espionage. By contrast, the chip design step (top row) has high importance in terms of revenue capture and spill-over benefits. This is reflected in Chinese authorities' support for chip design activity, hoping to use this as a revenue engine to drive growth of China's wider semiconductor industry and to support development of end-user industries. As a final example, chokepoints (sixth column) represent concentrated market share that can be 'weaponised' by governments of the states where the dominant firms are headquartered. EDA, SME, and cutting-edge fabrication have been weaponised by the United States against China through export controls, given US firms' dominance in the first two value chain steps and their importance to the third one (meaning that foreign leaders in cutting-edge fabrication like TSMC and Samsung are exposed to US jurisdiction under these controls).

Lee and Kleinhans (2021a) discuss how Figure 3.1's map of national interest relates to China's position in the different value chain steps. For instance, despite the high importance of cutting-edge fabrication, public Chinese policy and investment strategy has recently placed less emphasis on this value chain step. This reflects the difficulty of progress given high barriers to entry, and the constraints imposed by US export controls and pressure on allied countries to deny Chinese firms the necessary SME. However, the Party-State has been promoting R&D for component technologies for the requisite SME, as described below regarding photolithography.

Additionally, Figure 3.1 does not capture two significant variables. First, it is a frictionless model that does not account for perverse incentives and other factors that distort centralised policy implementation. For example, links between the national leading small group for the semiconductor sector and provincial equivalents are opaque. Official decisions on investments and administrative approvals in this sector still tend to be driven by bureaucratic incentives, rather than by industry advice and market realities (Fuller 2019; Randall 2022). Despite the importance of EDA and SME as chokepoints, and their promotion in official policy, these value chain steps have been neglected by the state's chief investment vehicle for the industry, the 'Big Fund' (discussed further below). Instead, the Big Fund has directed investments to expansion of fabrication capacity and the memory chip sector, where large capital investments can boost local economic growth figures and the investments are (by contrast with long term R&D efforts) more likely to generate short term profits.

Another variable not reflected in Figure 3.1 is the impact of transformative technological changes. The semiconductor value chain is relatively mature, and consistent with asymmetrical globalisation, it holds great inertia against actors changing their roles. However, the rising difficulty of further IC miniaturisation may herald the impending end of 'Moore's Law,' the decades-old

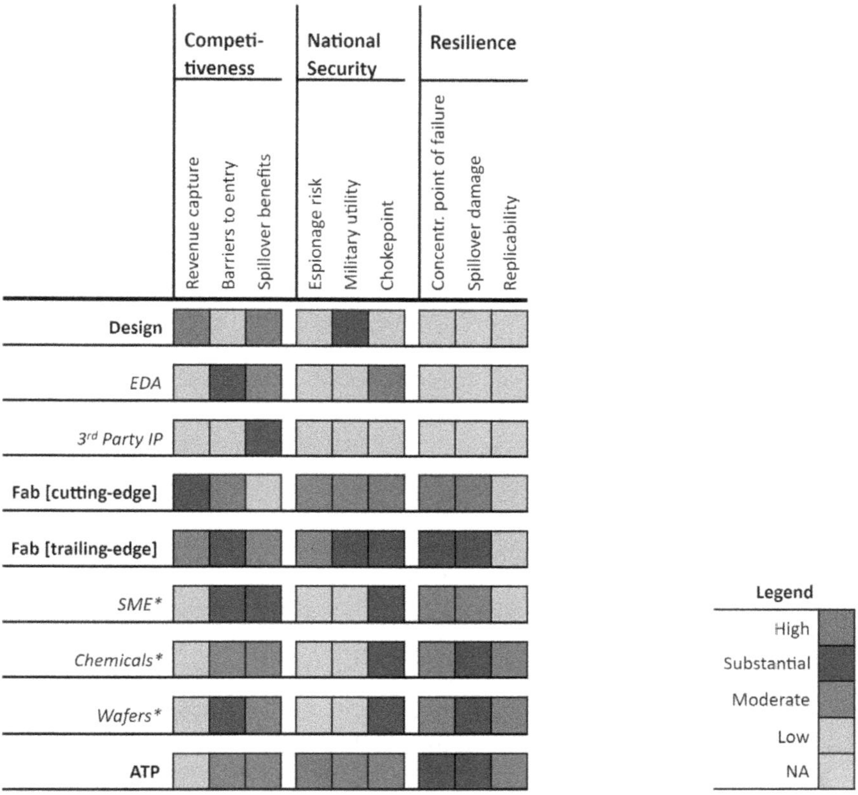


Process steps are in **bold**, inputs are in *italic*.

* The **barriers to entry** for *SME*, *chemicals* and *wafers* can be considered one level higher than listed in the matrix if used for cutting-edge wafer fabrication. As an example, while barriers to entry for chemicals suppliers are **moderate**, they are **substantial** for chemicals needed for cutting-edge wafer fabrication.

**Figure 3.1. Semiconductor Value Chain Steps Mapped Against National Interest Criteria**

Source: John Lee and Jan-Peter Kleinhans, "Mapping China's Semiconductor Ecosystem in Global Context: Strategic Dimensions and Conclusions," 30 June, 2021 Mercator Institute for China Studies and Stiftung Neue Verantwortung, p. 11.

observation that the number of transistors in a dense IC doubles around every two years, with concomitant increases in computing power. Sustaining such increases into the future will likely require new technical approaches. Combined with the imperative to circumvent chokepoints in the extant value chain structure represented in Figure 3.1, this has led Chinese industry and policymakers to focus on technical progress along pathways with significant technical development potential.

One example is ICs based on compound semiconductor materials, which have different electrical properties, rather than on the standard silicon. Compound semiconductors are now prioritised in numerous province and municipal-level government policies, notably by the Shanghai government, which hopes to create a 'Silicon Carbide Valley' (*Yicai* 2021). In turn this is driving behaviour by Chinese firms, including foreign acquisition activities, such as the recently blocked takeover of a UK facility by the Dutch subsidiary of a Chinese company (Guardian 2022). The utility of compound semiconductors for power management means that China's concentration of electric vehicle and other electronics manufacturing industries provides a strong source of supporting demand.

Another area that has received much attention is 'chiplets,' pre-developed subcomponents that support modular IC design, and potentially thereby achievement of more computing power using older generation chips. The CEO of Verisilicon, China's leading third-party semiconductor IP provider, has advocated building up a 'strategic stock' of chiplets to circumvent chokepoints in cutting-edge fabrication for advanced processors (*South China Morning Post* 2022). In early 2022, Chinese authorities were soliciting comments on a newly developed domestic standard for chiplet interfaces (CESA 2022). Chinese firms are members of the international industry consortium for chiplet standardisation, UCIe (BusinessWire 2022), but the scope of recent US export controls targeting China's semiconductor industry may force it onto its own pathway for chiplet technology.

Both the above examples must be qualified by recognising that these technologies for the time being remain path dependent, rather than transformative. While they may enable China to promote certain industry goals or mitigate the effects of chokepoint weaponisation, they remain constrained by the features of the semiconductor value chain represented in Figure 3.1. The prospects for a true paradigm shift in semiconductor technology remain opaque at best over a foreseeable timeframe, before even considering the question of whether China is the economy best placed to make or exploit such a hypothetical technological transformation.

## THE SEMICONDUCTOR ECOSYSTEM WITHIN CHINA

In 2014, China adopted its most recent dedicated industrial plan for the semiconductor industry (State Council 2014). This established a national leading small group to guide policy for the IC industry, as well as the National IC Industry Investment Fund ('Big Fund'). The Big Fund is a representative type of the so-called 'government guidance fund,' an institutional model for market-oriented 'grand steerage' that Chinese authorities have increasingly turned to over the last decade (Naughton 2022, 108; CSET 2021). These two institutions play key roles within the ecosystem of state, nonstate and mixed actors shaping activity in China's semiconductor industry (Figure 3.2).

## 'Grand Steerage'

'Leading small groups' (LSGs) in China bring together senior officials from different agencies with the aim of overcoming bureaucratic stove piping, inertia, and turf wars. At its establishment, the National IC LSG's deputy director was the head of the Ministry of Industry and Information Technology (MIIT), which leads development of China's digital technology-based sectors. It was

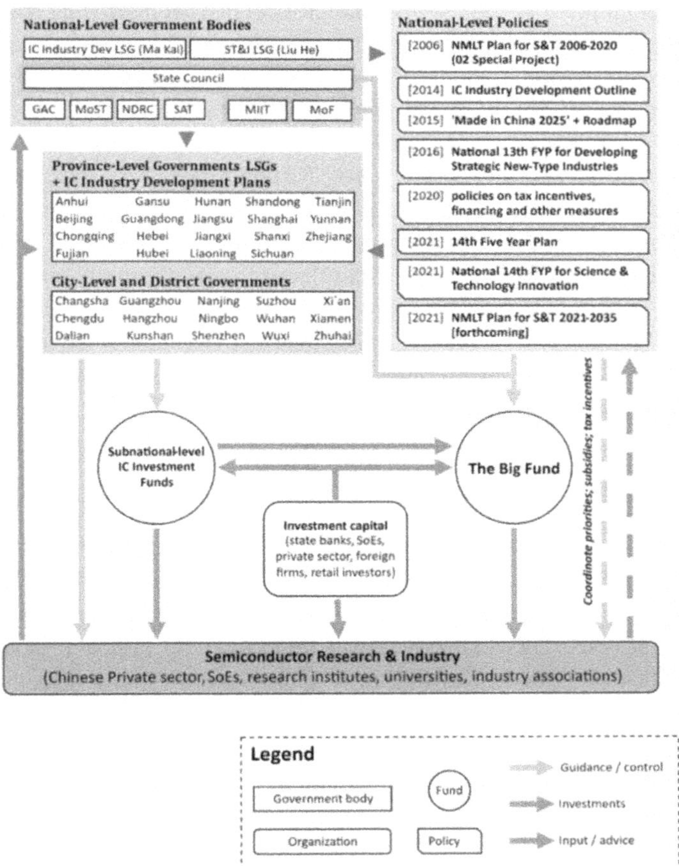


**Figure 3.2. China's Semiconductor Ecosystem**

Source: John Lee and Jan-Peter Kleinhans, Mapping China's semiconductor ecosystem in global context: Strategic dimensions and conclusions' 30 June, 2021 Mercator Institute for China Studies and Stiftung Neue Verantwortung, p. 16.

also advised by an 'A-Team' of experts on the semiconductor sector drawn from government, research institutions, industry, and investment funds (Ernst 2016, 7). Media reporting in 2021 suggested that that another LSG for 'reform of the national S&T system and building an innovation system' was also playing a key role in semiconductor policy. This would align with reorientation from a 'fast follower' approach to one focused on technology breakthroughs and leapfrogging (State Council 2021).

In addition to MIIT and the macroeconomic planning agency NDRC, other national agencies such as the Ministry of Finance (MoF), State Taxation Administration and General Customs Administration have been involved in issuing a range of supporting policies for the semiconductor industry as indicated in Figure 2. For instance, the latter three agencies in 2021 adopted a policy of exempting certain semiconductor-related technologies from import duties, in cases where domestic options are not available or cannot deliver the required performance (Gov.cn 2021).

Another key agency is the Ministry of Science and Technology (MoST), although criticism of bureaucratic influence over R&D funding have provoked efforts to curb MoST's role (Zhou, Lazonick, and Sun, 2016). Alongside MoF, NDRC and line ministries, MoST has played a lead role in coordinating the S&T development mega-projects specified in the 2006 NMLTP, including one for 'IC Manufacturing Equipment and Complete Technologies' (the '02 Special Project'). Running to 2020 in parallel with the NMLTP, this aimed to indigenise production of critical SME by assigning R&D tasks to different institutions. This effort's success is considered below regarding photolithography.

## The Role of Sub-National Governments

China's industrial policy system is decentralised in implementation, with sub-national governments accounting for a much higher level of state expenditure than the global average (Kroeber 2016, 4). By mid-2021, many province and city-level governments had developed their own IC industry development plans or governmental LSGs (Figure 3.2). Some of these plans seem likely to result in duplicated effort and wastage of resources pursuing unrealistic goals (Lee and Kleinhans 2021a, 33). Others seem to have better prospects, notably the Shanghai government's IC development plan, based on co-locating successful Chinese firms from different value chain steps and a bonded (import duty-free) manufacturing zone (Lee and Kleinhans 2021a, 22–23, 42).

'Supply chain mapping' initiatives, sometimes involving compulsory coordination forums with designated lead firms, seem to be an increasing feature of local government involvement (e.g., Jiangmen Municipal Government 2021; Chongqing Municipal Government 2021). Such requirements, which imply disclosing proprietary information, could become a major disincentive to foreign firms participating in China's semiconductor sector. For comparison, a more limited exercise in supply chain information collection by the US government in 2021 provoked pushback from South Korea and Taiwan, both at industry and government level (Lee and Kleinhans 2021b, 20–21).

## 'Government Guidance Funds' and Broad-Based Investment

Much reform in China since the 1980s has been aimed at developing sources of innovation outside the state's centralised administration, while maintaining state capacity to 'steer' industrial development (Zhou, Lazonick, and Sun 2016). Like other 'government guidance funds,' the 'Big Fund' is overseen by state agencies, specifically MIIT and MoF (Gov.cn 2014). Most of its shareholders are state-owned enterprises and other 'government guidance funds,' and the bulk of its initial registered capital came from MoF and the state-owned China Development Bank (CDB). The Big Fund's managing entity, Sino-IC Capital, has been continuously run by ex-CDB executives (*Caixin* 2020). By mid-2020, 14 province-level governments had set up their own IC investment funds, accounting for some 300 billion RMB (US$45 billion) (Lee and Kleinhans 2021a, 14).

One objective for 'government guidance funds' may have been avoiding World Trade Organisation restrictions on direct subsidies (US delegation to WTO 2018). But their key role is to provide sources of capital for domestic firms, a basic requirement for technological upgrading (Lin and Zhang 2019). With firms from advanced economies capturing 90 percent of revenues in the global semiconductor industry, Chinese firms were never going to become competitive through self-financing. State-linked funds aim to capitalise development of evolving strategic priorities, with the Big Fund's Phase 1 investments directed at fabrication and manufacturing-related categories of SME and materials, while Phase 2 pivoted to 'downstream investments'—for example, in compound semiconductor applications—in the hope of dragging along development of upstream sectors. The existence and investment choices of state-linked funds also incentivise private industry and venture capital actors, by signalling which areas are being favoured by government policy.

Media reports sometimes refer to a trillion renminbi (RMB) of planned Chinese investment into the semiconductor sector. This seems not to represent any officially published number, and likely derives from the stated aspiration for the Big Fund's Phase 2 (launched with registered capital of 204.5 billion RMB) to achieve a 1:5 multiplier effect in attracting further investment, the same ratio that Phase 1 is claimed to have achieved (Lee and Kleinhans 2021a, 14).

This approach of using state-linked funds to stimulate and lead commercial investment activity takes advantage of China's new public stock and private equity finance markets. In 2020 there were around 413 private equity deals in China's semiconductor sector worth around 140 billion renminbi (US$21 billion) (South China Morning Post 2021). Faced with a growing risk of losing access to imported technology, major actors like Huawei are investing in fellow Chinese companies that show prospects of developing domestic alternatives in key chokepoints. In EDA tools for instance, China's leading vendor (Empyrean) stated in its IPO prospectus a goal to develop a complete EDA ecosystem by 2025 (South China Morning Post 2022).

In 2021–2022, a government audit of the Big Fund, associated funds, and firms that received their investments was followed with investigations by the Party-State's' top disciplinary body into multiple executives at the Big Fund, Sino-IC Capital, and semiconductor firms (Financial Times 2022). While 'violations of law and discipline' have been cited, another likely reason for this crackdown is the continued weakness in SME and EDA that exposes wider Chinese industry to 'choking' by US export controls. Some of the Big Fund's investments appear to have been put to good use: memory chip maker YMTC, for example, was making sufficient progress with advanced memory that this technology was targeted by the October 2022 US export controls (despite being a commodity rather than a strategic asset). But the basic fact that despite such large investments, China remains so vulnerable to US leverage in this industry, shows the limitations of 'grand steerage' when faced with the market realities and technical complexities of the semiconductor value chain.

## China's Research-Industrial Complex

These complexities mean that Chinese technical progress has relied on R&D at state-resourced research institutions, especially given that many Chinese firms in this sector are relatively young and have small revenues. Much relevant Chinese IP is held by these state research institutions, including in emerging fields like compound semiconductors (KnowMade 2022). Staff at these institutions have significant influence over state policy and industry choices: for example, Wei Shaojun, director of Tsinghua University's Microelectronics Institute, is vice president of the China Semiconductor Industry Association, a Chinese delegate to the World Semiconductor Council (Triolo 2021) and a member of the National IC LSG's advisory 'A-Team' (Ernst 2016).

Private Chinese firms in the semiconductor sector typically spend a proportion of revenue on R&D comparable to foreign counterparts, but given their much lower revenues, this translates into much lower absolute spending (Randall 2021). With growing pressure on access to R&D partnerships abroad, including with foreign research institutions such as IMEC in Belgium (Bloomberg 2021), private Chinese firms are likely to increasingly depend on collaboration with state research institutions and on well-resourced state-owned enterprises, which will benefit from significant planned increases in the Chinese government's basic and applied research funding.

'Innovation' in the narrow sense is, however, only one side to a nation's technological progress. The other is its capacity to 'diffuse' new technology across society to actors who can use it effectively. With its command economy legacy and limited institutional reform since the 1980s, China still has a significant 'diffusion deficit' (Ding 2022; Zhou, Lazonick, and Sun 2016). The Chinese system has readily adopted some elements of its East Asian neighbours' formulae for technological upgrading, such as creation of enabling infrastructure in the form of 'industry park' type sectoral clusters. But China has been less successful in developing its own 'institutional foundations of the processes of technology leverage and diffusion management,' including the political space for industry to self-organise and have a genuine two-way interaction with government over strategic direction and priorities (Matthews and Cho 2009, 316, 319; Zhou, Lazonick, and Sun 2016, 49–50).

There is for example no Chinese equivalent to Taiwan's ITRI, a cross-sectoral public research and intermediary body tasked with identifying key technologies and diffusing them to the private sector, which directly birthed cutting-edge fabrication leader TSMC. And by contrast with the bottom-up formulation and implementation of S&T development in Taiwan, China still works with a top-down and relatively siloed system inherited from the Maoist era, albeit one that has undergone progressive rounds of reform to raise its effectiveness (Chang and Shih 2004; Zhou, Lazonick, and Sun 2016, 35–9). This reflects the Party-State's' recognition that while its S&T system has delivered success in natural state monopolies like a national space program and high-speed rail network, it has proved less effective in internationally competitive and market-driven sectors like semiconductors.

## China's Position in the Global Semiconductor Value Chain

The main features of China's aggregate position within the transnational semiconductor industry are fairly clear and have been identified by various researchers (Ernst 2016; Lee and Kleinhans 2021; Triolo 2021; Li 2021; Grimes and Du 2022). They can be summarised as follows:

- Chinese firms are now present throughout the value chain and are growing their capabilities and market share. Even faced with the extensive US export controls of October 2022, Chinese firms are probably already capable of supplying fabrication plants (fabs) short of the cutting-edge by plugging gaps in domestic production through open trade in noncontrolled items and black-market channels in controlled ones, and progressively import substituting even these items over the coming years.

- The most critical chokepoints for Chinese industry are EDA, cutting-edge fabrication, and key SME for equipping cutting-edge fabs, with the salient example being photolithography, as discussed below. EDA tool design is linked to fabrication, and so Chinese limitations in the latter will impede progress in the former. This may be mitigated by availability of pirated EDA software from the US industry leaders (Fuller 2021) and Chinese EDA vendors are relatively well positioned to expand their offerings, although the 'US persons' provisions of the October 2022 export controls have already led to some Chinese EDA startups losing key personnel.
- Chinese industry has been most successful in chip design and ATP, reflected in global market share in these value chain steps. But Chinese industry in general, and Chinese chip design firms especially, still rely on foreign-owned IP. Chinese firms are making progress with chip design based on RISC-V open-source ISA, which may provide the basis for a larger semiconductor IP ecosystem. But to date, RISC-V chips are yet to be adopted at scale even by the firms that designed them: for example, Alibaba still mainly uses US-designed chips for its cloud computing business. And ISAs are only one element of extensive third-party IP in the semiconductor industry.
- China also now accounts for a large and growing share of the world's trailing-edge fabrication capacity, thanks to state encouragement and the demand created by end-user industry concentration in China. This means that for less sophisticated chips used in a wide range of applications, China will be a major global supplier for the foreseeable future. This was acknowledged by one senior US official explaining the October 2022 export controls, which he described as not intended to stop the manufacture in China of chips to be used (for example) in car airbags (CNAS 2022).
- Despite critical self-assessments by Chinese industry of the nation's capacity to generate a sufficiently large semiconductor workforce, this is probably a lesser problem than the labour pool's practical industry experience (Lee and Kleinhans 2021a, 18–19; Fuller 2019). For this China has relied extensively on US, South Korean, and (especially) Taiwanese individuals, who have been instrumental in running operations and training up the domestic workforce. Greater controls by these foreign governments on their citizens' involvement in Chinese industry

may significantly slow the latter's progress, particularly in cutting-edge fabrication.

Perhaps the most critical constraint on China's progress toward cutting-edge fabrication, and thus domestically produced cutting-edge logic processors, is the SME category of photolithography. As noted above, the Dutch firm ASML monopolises the cutting-edge (EUV) of this technology, which it is not permitted to export to China under Dutch law. EUV machines are regarded as necessary for commercially viable production of ICs at 7nm (nanometers) fabrication process nodes or smaller. While China's leading fabrication firm SMIC has manufactured one chip design at a 7nm process node—news of which was reportedly a key trigger for the October 2022 US export controls (ChinaFile 2022)—SMIC probably achieved this with previous generation photolithography machines, under conditions that are unlikely to be commercially scaleable for a wide range of ICs.

Under the 02 Special Project, several Chinese institutions have been conducting R&D into requisite component systems for advanced photolithography. Media reporting in late 2022 indicated that these projects have reached a point where China's leading photolithography machine maker, SMEE, may soon prototype a photolithography machine for a 28nm process, capable of producing chips with adequate performance for a wide range of applications. SMIC operates 28nm and 14nm production lines using foreign-provided photolithography equipment, while one new Shenzhen-based fabrication firm linked to Huawei (which since 2020 has lost access to cutting-edge fabrication due to US export controls) reportedly aims to start 28nm production by 2025.

To put this in perspective, the October 2022 US export controls target Chinese fabrication capacity for logic chips at 14nm processes, two generations ahead of a 28nm process. The latter was brought into production in 2010 by fabrication leader TSMC, which in 2022 had a 5nm process in production and was about to introduce a 3nm process. Once SMEE produces a 28nm prototype, achieving commercial viability with this machine will present further challenges. And the technical leap in photolithography from 14nm to 7nm is immense, with R&D for some components still reportedly a significant distance from providing the basis for even a prototype EUV machine. Even with successful development of the most sophisticated components, replicating within China ASML's vast network of suppliers—some five thousand, according to recent annual reports—will remain a formidable task.

## CONCLUSION: CHINA'S PROSPECTS IN SEMICONDUCTORS AND STRUCTURAL INTERNATIONAL POWER

As a foundational technology, semiconductors are critical to China's development beyond a 'partial power' (Shambaugh 2013) on the global stage. The October 2022 US export controls are expressly motivated by fear of China's potential to deploy and employ semiconductor-based technologies like artificial intelligence more effectively than the United States. These measures have sprung from a policy debate fixated on a technology 'arms race,' and the need to pre-empt China reaching a 'tipping point' of technological capability beyond which it will inevitably outcompete the United States.

This chapter has shown a large gap between such concerns and the realities of China's place in the global semiconductor value chain. In this industry, China is dominant nowhere and has limited prospects to substitute key foreign dependencies in the short term, let alone to close the gap with foreign industry leaders at the technological frontier. Not only is the industry's structure highly unfavourable in many ways to Chinese firm growing their market share or technical capabilities, but China's own development ecosystem may hamper more than help, especially in view of growing constraints on the international access through which existing Chinese success has been built.

State influence remains strong enough to significantly distort market outcomes in inefficient ways (Fuller 2019), with state-backed firms collecting an estimated 60 percent of the semiconductor industry subsidies spent in 2020 (Nikkei Asia 2020). Xi's centralisation of authority, and the reassertion of ideological orthodoxy, is increasingly squeezing out the flexibility and experimentation in bureaucratic decision-making that has been crucial to China's past successes with industrial policy.

Despite rhetorical commitment to 'enterprises as the main locus of technological innovation' (Xinhua 2022), it is hard to characterise the Chinese system as emulating the other East Asian 'technology Tigers' in favouring entrepreneurship, rather than constraining it (Matthews and Cho 2009, 317). Chinese authorities still tend to favour SOEs or firms spun off from state institutions over the independent private sector, despite the former's mediocre performance and unimpressive track record for absorbing technology or dynamism in employing it commercially (Fuller 2019).

China's own public debate is cognisant of this technology 'diffusion deficit' (Ding 2022). But it is unclear whether this cognisance has reached the system's apex. A recent policy statement for 'key core technologies' like semiconductors (Xinhua 2022) emphasises tighter control by the Party centre through an 'authoritative decision-making command system,' and a 'combination of active government with an effective market.' This does not suggest a priority on rapid technological diffusion to industry actors given decision-making autonomy. Even where state-led policy has been most successful, in expansion of trailing-edge fabrication, it is unclear whether the state has focused enough on obtaining the human capital to effectively run all these new facilities (Fuller 2019).

The Chinese system is clearly capable of developing advanced technologies. Less clear is the system's capacity to employ and scale up these technologies in commercially viable and globally competitive ways, as achieved by semiconductor industry leaders like TSMC and ASML. The United States tolerated the rise of such foreign industry leaders because the countries involved were security allies, although in Japan's case tensions were eventually addressed through a bilateral agreement designed to protect the US semiconductor industry (Irwin 1994). China, by contrast, is trying to claw its way up in a mature industry dominated by US allies, in the face of committed US hostility.

Nonetheless, other features of the global semiconductor industry are more favourable to Chinese prospects. Many of the constituent technologies are progressing along the sort of clear developmental trajectories that provided 'leverage' for successful past cases of East Asian state-led industrial catch-up (Matthew and Cho 2009). Despite the expected pending end of Moore's Law, key semiconductor-related technologies are still advancing according to generally known developmental roadmaps that China can follow, with photolithography providing a case in point.

Experience in other industries also suggests caution in judging the future outcomes of Chinese state-led intervention by its past performance. In wireless telecoms for instance, despite limited adoption of China's indigenous 3G standard, the development experience helped Chinese firms improve their IP position in global 4G and 5G standards and translate this into global commercial success (Zhou, Lazonick, and Sun 2016; Malkin 2022). This contrasts, for example, with the experience of Japan, where isolated industrial development and state failure to create domestic markets resulted in technological 'leading without followers' (Kushida 2011; Lee 2020, 8).

The concentration in China of electronics manufacturing and emerging sectors like electric vehicles still exerts great attraction on foreign industry leaders (Lee 2021). Competition among foreign firms incentivises them to remain in China (Grimes and Sun 2016) for access to its enormous fast-growing markets, skilled labour pool, extensive infrastructure, and ecosystem of supplier firms, despite the *growing* difficulties and political risks of doing business there. This has been recently highlighted, for example, by leading German firms' announcement of major new investments and partnerships in China, and their CEOs' public intervention to advocate continued economic integration (Edge 2022). The reluctance of the Netherlands, Japan, and South Korea to replicate the October 2022 US export controls, despite pressure from Washington, shows how China's extant position in the semiconductor sector and end-user industries affects international power relations.

The partial success of China's semiconductor ecosystem, and its limited prospects for closing the gap with the US-allied community in most of the industry's commanding heights, suggests that China's further progress in this field will append rather than supplant US dominance (Malkin 2022). China's growing presence in chip design, trailing-edge fabrication and ATP already has clear implications for the resilience of the global supply of semiconductors and the dependencies of foreign industry actors (Lee and Kleinhans 2021b). Elsewhere, Chinese industry will suffer from US export controls, especially in its international competitiveness. But while US controls may achieve their stated goal of precluding Chinese technological leadership, they are unlikely to stop China's accumulation of structural power in the global economy, as embedded in technological networks. This trend may not deliver on the Party-State's' ambitions for semiconductors and other advanced technologies, but it is growing the resources and international leverage available to Beijing.

The realities of the semiconductor value chain mean that every country involved is a 'partial power,' including the United States itself. In such a key technology, even limited success in breaking the mould of 'asymmetrical globalisation' enhances China's power in the international system and its capture of gains from trade. This appears increasingly to be a model for other large developing countries, notably India and Indonesia. Even the advanced economies are now adopting ambitious industrial policy for semiconductors, recognising their implications for the distribution of wealth and power. But policymakers should be wary of making choices that harm their own economies, by clamping down on international trade and globalised innovation simply because this also benefits countries that are perceived as strategic rivals. Despite the concentrated character and strategic importance of the semiconductor industry, it presents a case for managing economic interdependence with China, rather than seeking China's technological containment in absolute terms.

## REFERENCES

Antipolis, Sophia. 2022. Silicon Carbide (SiC) patents support the emergence of a complete domestic supply chain in China. *KnowMade Patent & Technology Intelligence, 20 January*. Available from: https://www.knowmade.com/technology-news/semiconductor-news/power-electronics-devices-news/silicon-carbide-sic

-patents-support-the-emergence-of-a-complete-domestic-supply-chain-in-china [9 December 2022].

Austin, G. 2014. *Cyber Policy in China.* Cambridge: Polity Press.

Bresnick, S. and Sher, N. 2022. New Export Controls on Chinese Semiconductors May Prove Self-Defeating. China File, 16 September. Available from: https://www.chinafile.com/reporting-opinion/viewpoint/new-export-controls-chinese-semiconductors-may-prove-self-defeating [9 December 2022].

Bureau of Industry and Security, United States Department of Commerce. 2022. Implementation of Additional Export Controls: Certain Advanced Computing and Semiconductor Manufacturing Items; Supercomputer and Semiconductor End Use; Entity List Modification. Federal Register. 7 October. Available from: https://public-inspection.federalregister.gov/2022-21658.pdf [9 December 2022].

Business Wire. 2022. Leaders in Semiconductors, Packaging, IP Suppliers, Foundries, and Cloud Service Providers Join Forces to Standardize Chiplet Ecosystem. *Business Wire.* 2 March. Available from: https://www.businesswire.com/news/home/20220302005254/en/Leaders-in-Semiconductors-Packaging-IP-Suppliers-Foundries-and-Cloud-Service-Providers-Join-Forces-to-Standardize-Chiplet-Ecosystem [9 December 2022].

Caixin. 2020. Head of Firm That Runs National Chipmaking Investment Fund Steps Down. Caixin, 23 December. Available from: https://www.caixinglobal.com/2020-12-23/head-of-firm-that-runs-national-chipmaking-investment-fund-steps-down-101642641.html [9 December 2022].

Cheung, Tai Ming. 2018. The rise of China as a cybersecurity industrial power: balancing national security, geopolitical, and development priorities. *Journal of Cyber Policy* 3(3), 306–26.

*China-cer.com.cn.* 2022. Guanyu jianquan shehuizhuyi shichang jingji tiaojianxia guanjian hexin jishu zhengguan xingxing juguo tizhi de yijian ('Opinions on Improving the New National System for Tackling Key Core Technologies under the Conditions of Socialist Market Economy'). *China-cer.com.cn*, 8 September. Available from: http://www.china-cer.com.cn/zhengcefagui/2022090821090.html [9 December 2022].

Coe, Nei, Hess, Martin, Yeung, Henry Wai-Chung, Dicken, Peter and Henderson, Jeffrey. 2004. 'Globalizing' Regional Development: A Global Production Networks Perspective. Transactions of the Institute of British Geographers 29(4), 468–84.

Crawford, Alan. 2021. The U.S.-China Tech Conflict Front Line Goes Through Belgium. *Bloomberg*, 13 July. Available from: https://www.bloomberg.com/news/features/2021-07-13/u-s-and-china-fix-their-sights-on-world-s-top-chip-research-center [9 December 2022].

Deloitte. 2022. 2022 semiconductor industry outlook. Available from: https://www2.deloitte.com/us/en/pages/technology-media-and-telecommunications/articles/semiconductor-industry-outlook.html [9 December 2022].

Eckl-Dorna, W. 2022. German CEOs warn China exit would threaten growth, prosperity. *The Edge Markets*, 10 November. Available from: https://www.theedgemarkets.com/article/german-ceos-warn-china-exit-would-threaten-growth-prosperity [9 December 2022].

The Economic Times. 2022. India to invest $30 billion in tech sector, semiconductor supply chain: report. The Economic Times, 16 June. Available from: https://economictimes.indiatimes.com/tech/technology/india-to-invest-30-billion-in-tech-sector-semiconductor-supply-chain-report/articleshow/92243988.cms [9 December 2022].

Ernst, D. 2016. The Information Technology Agreement, Manufacturing and Innovation—China's and India's Contrasting Experiences. East-West Center, Honolulu and Centre for International Governance Innovation, Waterloo. Available from: https://www.eastwestcenter.org/sites/default/files/filemanager/pubs/pdfs/6-1Ernst-20160925.pdf [9 December 2022].

Ernst, Dieter, and Kim, Linsu. 2002. *Global production networks, knowledge diffusion, and local capability formation*. Research Policy 31, 1417–29.

Financial Times. 2022. China's Big Fund corruption probe casts shadow over chip sector. *Financial Times*, 29 September. Available from: https://www.ft.com/content/8358e81b-f4e7-4bad-bc08-19a77035e1b4 [9 December 2022].

Fuller, Douglas. 2014. Chip design in China and India: Multinationals, industry structure and development outcomes in the integrated circuit industry. *Technological Forecasting and Social Change* 81, 1–10.

Fuller, D. 2019. Growth, Upgrading, and Limited Catch-Up in China's Semiconductor Industry. In Brandt, L. and Rawski, T., eds. *Policy, Regulation and Innovation in China's Electricity and Telecom Industries*. Cambridge: Cambridge University Press.

Fuller, D. 2016. *Paper Tigers, Hidden Dragons: Firms and the Political Economy of China's Technological Development*. Oxford: Oxford University Press.

Great Wall Glory Securities. 2022. Hangye chuyu guochan tidai chuqi, 2022nian shebei xuqiu qiangjing ('The Industry remains in an early phase of import substitution, equipment demand is strong in 2022'). *Eastmonye.com*, 19 May. Available from: https://data.eastmoney.com/report/zw_industry.jshtml?infocode=AP202205191566491722 [9 December 2022].

Grimes, Seamus and Sun, Yutao. 2016. China's evolving role in Apple's global value chain. Area Development and Policy 1(1), 94–112.

Grimes, Seamus and Du, Debin. 2022. China's emerging role in the global semiconductor value chain. *Telecommunications Policy* 46(2), 1–14.

Huang, R. and Henderson, J. 2021. Is There a Method Behind China's Tech Crackdown Madness? MacroPolo, 21 October. Available from: https://macropolo.org/china-tech-crackdown-software-hardware/?rp=e [9 December 2022].

Hwang, Hye-Ran and Choung, Jae-Young. 2014. The Co-evolution of Technology and Institutions in the Catch-up Process: The Case of the Semiconductor Industry in Korea and Taiwan. *The Journal of Development Studies* 50(9), 1240–60.

IC Insights. 2021. Cited in: Sales of Logic ICs Account for Largest Share of China's IC Market in 2020. *Design and Reuse*, 18 February. Available from: https://www.design-reuse.com/news/49515/2020-china-ic-market-by-product-type.html [9 December 2022].

Irwin, D. 1996. The U.S.-Japan Semiconductor Trade Conflict. In Krueger, A. ed. *The Political Economy of Trade Protection*. Chicago: University of Chicago Press.

Jolly, J., and Badshah, N. 2022. British government blocks takeover of Welsh semiconductor producer. *The Guardian*, 16 November. Available from: https://www.theguardian.com/technology/2022/nov/16/british-government-blocks-takeover-of-welsh-semiconductor-producer [9 December 2022].

Kleinhans, J. 2022. U.S.–China Competition in Global Supply Chains. Testimony before the U.S.–China Economic and Security Review Commission. 9 June. Available from: https://www.uscc.gov/sites/default/files/2022-06/Jan-Peter_Kleinhans_Testimony.pdf [9 December 2022].

Koc, C. and Wu, D. 2022. Dutch Minister Says US Can't Dictate Approach to China Exports. Bloomberg, 18 November. Available from: https://www.bloomberg.com/news/articles/2022-11-18/dutch-minister-says-us-can-t-dictate-approach-to-china-exports [9 December 2022].

Kroeber, Arthur. 2016. *China's Economy: What Everyone Needs to Know*. Oxford: Oxford University Press.

Kushida, Kenji. 2011. Leading without Followers: How Politics and Market Dynamics Trapped Innovations in Japan's Domestic 'Galapagos' Telecommunications Sector. *Journal of Industry, Competition and Trade* 11: 279–307.

Lee, J. and Kleinhans, J. 2020. Taiwan, Chips, and Geopolitics: Part 1. *The Diplomat*, 10 December. Available from: https://thediplomat.com/2020/12/taiwan-chips-and-geopolitics-part-1/ [9 December 2022].

Lee, J. 2021. The Connection of Everything: China and the Internet of Things. Mercator Institute for China Studies. 24 June. Available from: https://merics.org/en/report/connection-everything-china-and-internet-things [9 December 2022].

Lee, J. and Kleinhans, J. 2021a. Mapping China's semiconductor ecosystem in global context: Strategic dimensions and conclusions. Mercator Institute for China Studies and Stiftung Neue Verantwortung. 30 June. Available from: https://merics.org/en/report/mapping-chinas-semiconductor-ecosystem-global-context-strategic-dimensions-and-conclusions [9 December 2022].

Lee, J. and Kleinhans, J. 2021b. China's rise in semiconductors and Europe: Recommendations for policy makers. Mercator Institute for China Studies and Stiftung Neue Verantwortung. 8 December. Available from: https://merics.org/en/report/chinas-rise-semiconductors-and-europe-recommendations-policy-makers [9 December 2022].

Li, Y. The Semiconductor Industry: A Strategic Look at China's Supply Chain. In F. Spigarelli and J. McIntyre, eds. *The New Chinese Dream: Industrial Transition in the Post-Pandemic Era*. Cham: Palgrave Macmillan.

Lin, J. and Zhang, J. 2019. China: Learning to Catch up in a Globalized World. In Oqubay, A. and Ohno, K., eds. *How Nations Learn: Technological Learning, Industrial Policy, and Catch-up*. Oxford: Oxford University Press.

Matthews, J. and Cho, D. 2009. *Tiger Technology: The Creation of a Semiconductor Industry in East Asia*. Cambridge: Cambridge University Press.

Mercator Institute for China Studies. 2021. 'The CCP's next century: expanding economic control, digital governance and national security.' 15 June. Available from: https://merics.org/en/report/ccps-next-century-expanding-economic-control-digital-governance-and-national-security [9 December 2022].

Ministry of Finance, General Administration of Customs and General Administration of Taxation. Caizhengbu, Haiguan Zongshu, Shuiwu Zongju, Guanyu zhichi jicheng dianlu chanye he ruanjian chanye fazhan jinkou shuishou zhence de tognzhi ('Circular on Import Taxation Policies to Support the Development of Integrated Circuit Industry and Software Industry'). Available from: http://www.gov.cn/zhengce/zhengceku/2021-03/29/content_5596564.htm [9 December 2022].

Malkin, Anton. 2020. The made in China challenge to US structural power: industrial policy, intellectual property, and multinational corporations. *Review of International Political Economy*, 1–33.

McCarthy, D. 2015. *Power, Information Technology, and International Relations Theory: The Power and Politics of US Foreign Policy and Internet*. London: Palgrave Macmillan.

Na, E. 2022. South Korea caught in the middle of US-China chip war, but American export control requests unlikely. South China Morning Post, 14 November. Available from: https://www.scmp.com/economy/china-economy/article/3199299/south-korea-caught-middle-us-china-chip-war-american-export-control-requests-unlikely_[9 December 2022].

Naughton, B. 2. 2022. Grand Steerage as the New Paradigm for State-Economy Relations. In F. Pieke and B. Hofman, eds. *CPC Futures: The New Era of Socialism with Chinese Characteristics*. Singapore: NUS Press, 105–12.

Office of the United States Trade Representative (OUSTR). 2022. Remarks by Ambassador Katherine Tai at the Roosevelt Institute's Progressive Industrial Policy Conference. OUSTR. Available from: https://ustr.gov/about-us/policy-offices/press-office/speeches-and-remarks/2022/october/remarks-ambassador-katherine-tai-roosevelt-institutes-progressive-industrial-policy-conference [9 December 2022].

Orr, G. and Thomas, C. 2014. Semiconductors in China: Brave new world or same old story? McKinsey. Available from: https://www.mckinsey.com/~/media/McKinsey/dotcom/client_service/Semiconductors/PDFs/Semiconductors_in_China_Brave_new_world_or_same_old_story [9 December 2022].

Oxford Analytica. 2022. New plan will rebalance China's digital growth. Oxford Analytica. 4 February. Available from: https://www.emerald.com/insight/content/doi/10.1108/OXAN-DB267130/full/html/ [9 December 2022].

Pan, C. 2021. China's state semiconductor fund trims holdings in SMIC as Beijing focuses on next wave of development. *South China Morning Post*, 16 April. Available from: https://www.scmp.com/tech/big-tech/article/3129892/chinas-state-semiconductor-fund-trims-holdings-smic-beijing-focuses [9 December 2022].

Pan, C. 2022. Tech war: China experts at odds over role of 'chiplets' in helping achieve goal of semiconductor self-sufficiency. *South China Morning Post*, 3 June. Available from: https://www.scmp.com/tech/big-tech/article/3180015/tech-war-china-experts-odds-over-role-chiplets-helping-achieve-goal [9 December 2022].

Pan, C. 2022. Chinese firm aiming to break US dominance in chip design software gets IPO approval at home. *South China Morning Post*, 24 June. Available from: https://www.scmp.com/tech/tech-war/article/3182988/chinese-firm-aiming-break-us-dominance-chip-design-software-gets-ipo [9 December 2022].

Poon, D. 2018. China broke the rules of global trade—but for good reason. *South China Morning Post*, 21 June. Available from: https://www.scmp.com/comment/insight-opinion/united-states/article/2151688/china-broke-rules-global-trade-good-reason [9 December 2022].

Randall, S. 2021. Are Chinese chipmakers spending enough on R&D? TechNode, 10 September. Available from: https://technode.com/2021/09/10/silicon-are-chinese-chipmakers-spending-enough-on-rd/ [9 December 2022].

Randall, S. 2022. Why is China investigating the state-backed semiconductor "Big Fund"? TechNode. 12 August. Available from: https://technode.com/2022/08/12/silicon-why-is-china-investigating-the-state-backed-semiconductor-big-fund/ [9 December 2022].

Rasser, M. 2022. 'A Conversation with Under Secretary of Commerce Alan F. Estevez' (transcript). Available from: https://www.cnas.org/publications/transcript/a-conversation-with-under-secretary-of-commerce-alan-f-estevez [9 December 2022].

Rho, Sungho, and Kim, Yongshin. 2021. Sectoral Divergence of Industrial Catch-Up in China's Loosely Coupled System: A Comparative Study of FPD and IC Manufacturing Industries. *Pacific Focus* 36(3), 512–43.

Shambaugh, D. 2014. *China Goes Global: The Partial Power.* Oxford: Oxford University Press.

SIA: Semiconductor Industry Association. 2021. 2021 Factbook. Available from: https://www.semiconductors.org/wp-content/uploads/2021/05/2021-SIA-Factbook-FINAL1.pdf [9 December 2022].

State Council. 2021. Zhonghuaremingongheguo Guomin jingji he shehui fazhan dishisige wunian guihua he 2035 nian yuanjing mubiao gangyao ('14th Five Year Plan of the People's Republic of China for the nation's social and economic development, and outline of long-term goals to 2035'). *Gov.cn*, 27 December. Available from: http://www.gov.cn/xinwen/2021-03/13/content_5592681.htm [9 December 2022].

State Council. 2021. 'Shisiwu' Guojia Xinxihua Guihua ('14th Five Year Plan for National Informatisation'). Cyberspace Administration of China, 13 March. Available from: http://www.cac.gov.cn/2021-12/27/c_1642205314518676.htm [9 December 2022].

Thomas, C. 2021. Lagging but motivated: the state of China's Semiconductor Industry. Brookings Institution, 7 January. Available from: https://www.brookings.edu/techstream/lagging-but-motivated-the-state-of-chinas-semiconductor-industry/ [9 December 2022].

To, Y. 2022. *Contested Development in China's Transition to an Innovation Driven Economy*. Abingdon: Routledge.

Triolo, P. 2021. The Future of China's Semiconductor Industry. *American Affairs* V(I). Available from: https://americanaffairsjournal.org/2021/02/the-future-of-chinas-semiconductor-industry/ [9 December 2022].

UNCTAD: United Nations Conference on Trade and Development. 2018. Trade and Development Report 2018. Available from: https://unctad.org/webflyer/trade-and-development-report-2018 [9 December 2022].

US Delegation to the World Trade Organisation. 2018. China's Trade-Disruptive Economic Model. Communication from the United States, 11 July. Available from: https://docs.wto.org/dol2fe/Pages/SS/directdoc.aspx?filename=q:/WT/GC/W745.pdf [9 December 2022].

Verwey, J. 2019. Chinese Semiconductor Industrial Policy: Past and Present. Chinese Semiconductor Industrial Policy: Past and Present. *United States International Trade Commission Journal of International Commerce and Economics*, 1–29.

Xi Jinping. 2020. Guojia zhongchangqi jingjishehuifazhanzhanlue ruogan zhongda wenti ('Some important problems concerning the nation's medium and long-term Economic and Social Development Strategy'). *Qiushi*, 31 October. Available from: http://www.qstheory.cn/dukan/qs/2020-10/31/c_1126680390.htm [9 December 2022].

Xinhua. 2014. Guojia jicheng dianlu change touzi jijin zhengshi jianli ('National Integrated Circuit Industry Investment Fund formally established'). *Gov.cn*, 14 October. Available from: http://www.gov.cn/xinwen/2014-10/14/content_2764849.htm [9 December 2022].

Xinhua. 2014. Xi Jinping: Ba wo Guo cong 'Wangluo Daguo' jianshe chengwei 'Wangluo Qiangguo' ('Xi Jinping: Build the nation from a 'Cyber Great Power' into a 'Cyber Superpower"). *Xinhuanet.com*, 27 February. Available from: http://www.xinhuanet.com/politics/2014-02/27/c_119538788.htm [9 December 2022].

White House. 2022. Remarks by National Security Advisor Jake Sullivan at the Special Competitive Studies Project Global Emerging Technologies Summit. White House. 22 September. Available from: https://www.whitehouse.gov/briefing-room/speeches-remarks/2022/09/16/remarks-by-national-security-advisor-jake-sullivan-at-the-special-competitive-studies-project-global-emerging-technologies-summit/ [9 December 2022].

WSTS: World Semiconductor Trade Statistics. 2022. WSTS Semiconductor Market Forecast Spring 2022. 7 June. Available from: https://www.wsts.org/esraCMS/extension/media/f/WST/5550/WSTS_nr-2022_05.pdf [9 December 2022].

Yu, Z. 2021. Jianshe quanqiu yingxiangli "Dongfang Xingang"! Shanghai Lingang Xinpianqu fabu jichengdianlu chanye zhuanxiang guihua ("Build the 'Eastern Chip Gateway' with global influence! Shanghai Lingang New Area issues plan for integrated circuit industry and special projects"). Yicai.com. 3 March. Available from: https://www.yicai.com/news/100966145.html [9 December 2022].

Zhongguo Zhengfuwang. 2021. Liu He zhuchi zhaokai Guojia Keji Tizhi Gaige he Chuangxin Tixi Jianshe Lingdao Xiaozu dishibaci huiyi ('Liu He Chaired the 18th Meeting of the Leading Group on the Reform of the National Science and Technology System and the Construction of the Innovation System'). *Gov.cn*, 14 May. Available from: http://www.gov.cn/guowuyuan/2021-05/14/content_5606512.htm [9 December 2022].

Zhou, Y., Lazonick, W. and Sun, Y., eds. 2016. *China as an Innovation Nation*. Oxford: Oxford University Press.

# Chapter 4

# Fintech in China

## *Trading off Growth and Risk, Innovation, and Control*

Martin Chorzempa

Finance is the lifeblood of any economy, and Chinese financial technology or 'fintech' and its digital economy have boomed in a symbiotic relationship. The government's view of fintech has changed over time, but overall it is seen as a beneficial tool for contributing to China's economic and technological development. Technology adoption can make the financial system more efficient, lower costs, and serve more people, and China has had immense success thanks to the growth of its fintech sector. China's fintech sector, the largest in the world by far, has also brought beneficial competition with protected state monopolies that provided poor service and little innovation.

Fintech has also become a soft power asset for China, an area where it can claim to have 'leapfrogged' advanced economies to become the world leader. Fintech remade its backward cash-based financial system, forging instead a system designed for use with mobile phones and digital commerce. Payments in China with digital wallets are many times cheaper than those done with credit cards in the United States, where people still carry around plastic cards and sometimes even paper checks to make payments. Before the pandemic shut off most travel to China, foreigners marveled at the advanced state of Chinese mobile payments, which bankers describe as 'close to the end state' of what banking could look like in the future (Engen 2018). China's top fintech companies have invested billions in fintech firms around the world and expanded acceptance of Chinese payments apps to dozens of countries, leveraging their technology, capital, and expertise to make their mark outside of China. Many believe the digital RMB project from China's central bank could use technology to remake the way the RMB is used in international payments, reducing the leverage and sanction power the United States currently wields thanks to the dominant role of the dollar in global commerce and investment.

Yet, the rapid growth of innovative finance has not been an unalloyed good. It has also brought the risk of financial instability, the creation of new potentially unassailable monopolies in the form of 'super apps,' as well as data protection concerns. Authorities are now hoping to reduce risk and disruption without cutting off needed credit or the room for still needed financial innovation. They have largely been successful at reducing risk, but their hope to remake the fintech sector will need years to implement complex and sometimes mutually contradictory policy goals.

## FINTECH'S IMPORTANCE TO DIGITAL DEVELOPMENT

When China's largest internet firms were getting started in the early 2000s, China's cash-based payment system was a major impediment to their business. Unlike in the United States, digital advertising provided too few opportunities to raise revenue, so they needed to collect payment for digital goods directly from consumers to become profitable. The problem, however, was that payments through credit cards and other digital means taken for granted in advanced economies were inconvenient or not available. Tencent, one of China's internet giants with a focus on social media and gaming, struggled to collect small payments for in-game items or cheap subscriptions to digital services. 'Pony' Ma Huateng, Tencent's founder, recalled that 'almost none of China's young consumers had a credit card. They had to run to the post office to make a transfer, which few netizens were willing to do for a 10 RMB payment every month' (quoted in Wu 2017). It issued its own virtual currency, the Q coin, so that users could make one payment to Tencent with the clunky legacy financial system and then use a digital one with Q coins to make smaller purchases seamlessly and with no fees.

China's State Council recognised the problem. It issued opinions in 2005 to signal clear support for online payments to help electronic commerce 'change the way our economy grows and raise the quality and efficiency of citizens' economic activity' (State Council 2005). Though state-backed China Union Pay retained a monopoly on card payments, playing a role analogous to Mastercard or Visa in the United States, the government encouraged the development of private online payments options. China's central bank, the People's Bank of China (PBOC), left online payments free of regulation until 2010, seemingly a libertarian paradise in an authoritarian country, to 'create a relatively loose environment for the innovation and development of e-payment business' (PBOC 2005).

E-commerce, meanwhile, struggled with payments but also trust. Most transactions initially occurred between people in the same city, who inspected the goods and paid cash when they were deemed satisfactory, limiting its scope (Barnett and Lorentzen 2006). In the United States, this was less of a problem because credit cards contain consumer protections that ensure customers are not charged if the goods do not arrive. American e-commerce companies thus did not need to build their own payment tools, but Alipay helped solve this problem for Alibaba's Taobao consumer to consumer marketplace. Alipay started as an escrow service, guaranteeing buyers they would not be charged unless goods arrived in satisfactory condition while also guaranteeing merchants that they would be paid if they delivered.

The official government encouragement was crucial to providing political and regulatory space for entrepreneurs like Jack Ma, Alibaba's founder, to engage in this problem solving, though he claims to have moved forward with Alipay after assuring his staff that he would be the first if anyone had to go to jail (Ma 2018). Online payments started for online purchases but expanded to offline purchases, like taxis and in-person restaurant dining around 2014, which further digitised payments. More digital commerce created demand for more digital financial tools like mobile payments, generating data and the capability to generate insights from those data.

Both Alipay and Tencent's WeChat Pay, part of its new mobile chat app, made collecting digital money fast and cheap, averaging a fee of about 0.6 percent per transaction, many times cheaper than credit cards in the United States that can reach 2 to 3 percent. Since the systems were digital native, they were cheap to process and carried no fixed fee per payment, which makes it economical to send even payments of a few cents. Business models that thrive in China like tipping and paying per article for media would not be possible with credit cards. Digital payment then became infrastructure crucial for the development of China's broader digital economy, shaping the possibilities of business that can thrive online and off.

Credit is also essential to a modern economy, spurring consumption, entrepreneurship, and efficient functioning of business. Fundamentally, credit is an information issue. Credit providers must find and screen potential borrowers, for example with a credit score, and monitor them after the loan is given, for example to ensure the funds are not put into the cryptocurrency market instead of buying business supplies. Fintech's use of data could help improve lending by giving credit providers better information. Before fintech, household loans were mostly limited to purchases of housing and cars with large down payments. Borrower characteristics are less important when a defaulting borrower would just forfeit the collateral. Consumer credit is harder, and small business credit is a major challenge around the world, because small firms do not tend to have assets that can be put up as collateral.

The traditional way to evaluate such borrowers outside China is to look at their credit history, but China faced a chicken and egg problem because it historically had so little formal consumer and small business credit. Few lenders wanted to extend credit to the hundreds of millions of Chinese without credit histories, but unless that changed there was no way they could get a credit history needed to borrow. Fintech overcame the chicken and egg problem with big data sets generated by online payments and commerce to develop credit screening tools that allowed lenders to estimate even a first-time borrower's financial resources and risk level, all without costly human involvement in lending decisions. They could also generate credit histories quickly by lending small amounts to consumers very short term. Control of the payment system helped monitor what the borrowed funds were spent on. Ant Group, spun off from Alibaba, became one of China's largest lenders with this model built into Alipay.

For small business credit, e-commerce and running the payments system provided Alibaba and other firms like JD with real time information on an online business's health that they could use to control credit risk. Because the seller relies on the platform, they have a strong motivation to repay, lest they be kicked down the rankings or risk losing access to their revenues from Alipay.

## MAJOR PLAYERS

The shape of fintech in China is the result of competition between fintech firms, collaboration and competition with traditional financial institutions, and a regulatory role for the government that shapes the incentives and activities each can undertake. Governmental policy initially was a double-edged sword. Before fintech, the financial system implemented financial repression, in which regulatory barriers reserved most of finance, especially payments and banking, for state-owned companies. Chinese savers' money would then be channeled at low cost to state owned firms and the government to implement state priorities like infrastructure investment (Chorzempa 2022). Those barriers kept fintech from entering much of finance, and artificially constrained supply of credit and other financial services left an immense addressable market for innovators—if the government would permit them to compete.

Around 2012, market reform-minded officials like PBOC Governor Zhou Xiaochuan convinced the leadership that competition was needed to improve the financial system and support economic growth. Then-Premier Wen Jiabao said, 'We're dealing with the issue of getting private capital into the finance sector, essentially, that means we have to break up [the state banks'] monopoly' (Reuters 2012). Regulators then implemented the promised openings, permitting fintech firms to enter the wealth management, consumer lending, offline payments, banking, insurance, and other sensitive markets.

The most important fintech companies are Alibaba and Tencent. Alibaba's main strength is in e-commerce, while Tencent is primarily a gaming and social media company. Both had strong political patronage, technical expertise, and large bodies of users to which they could push financial services. Though their main lines of business did not overlap that much until the arrival of smartphones, mobile internet launched an era of platforms, in which each turned their main apps into an ever-larger bundle of services. One of the first competitive areas was so-called online to offline (O2O) services that could use smartphones to order in-person goods and services like Didi and Kuaidi (equivalent to Uber elsewhere). Alibaba and Tencent competed with a war of subsidies to get consumers and businesses to adopt QR code payments for such services, starting with taxis and, as figure 4.1 shows, becoming the main way Chinese paid for items online and off starting in 2016.

Tech firms competing achieved what over a decade of state monopoly, as China UnionPay cards issued by the state banks, could not: a wholesale shift from cash to digital payments. Alibaba affiliate Alipay had a 55 percent market share of the nonbank mobile payments market in 2020, and the latter 38 percent, together thus controlling 93 percent—an effective duopoly (Analysys 2020). Alipay had 711 million monthly active users in 2020, more than 60 percent of China's adult population (Alipay 2020), and WeChat Pay counts more than one billion users. UnionPay tried and failed to preserve its de facto

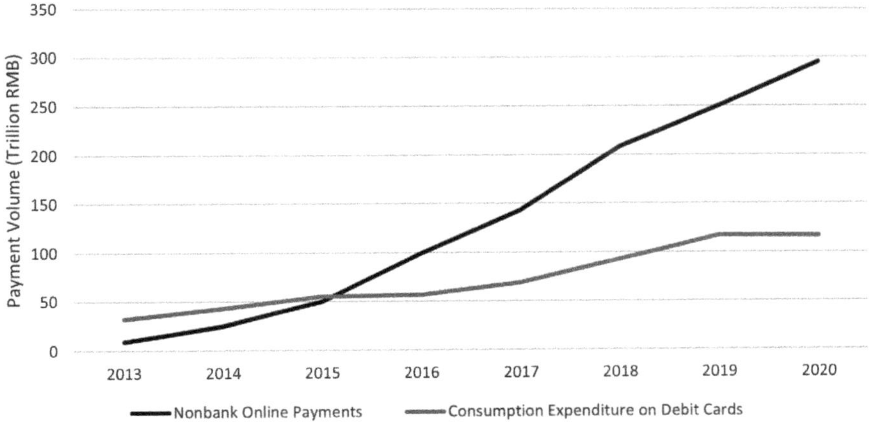


**Figure 4.1. How Chinese Citizens Pay for Items Online and Off**

Source: Author calculations based on People's Bank of China data.

monopoly on offline payments, but popular support and allies in government protected fintech's disruption—for the benefit of Chinese consumers.

Tech firms' platform model led to China's fintech unique strengths, which come from fusing finance with nonfinancial services and goods in a single application. Super apps have become more akin to operating systems on which a host of other apps, including mini apps, can build. Alipay and WeChat's control of payments allows them to provide access to their users and the ability to charge those users to third parties that can build their business on the Alibaba or WeChat platform.

ICBC, China's largest state-owned bank, partnered with Alipay around 2003 to process payments and hold customer funds safely, but once Alipay expanded beyond payments a decade later the relationships with banks became more tense. Banks hold strong political power since their executives have a place similar to vice ministers in the Party's Nomenklatura and have long been connected with implementing party policy (McGregor 2012). Alipay's disruption of financial repression through its launch of a money market fund in mid 2013 drained funds out of the banks and marked a new era of both economic competition and political competition, in which fintech firms would probe the limits of disruption authorities would condone. Later rules would force fintech firms into a more collaborative stance with the banks, helping them upgrade their IT systems and leveraging their data to help banks make loans instead of a focus on providing competing products.

Thus, regulators have played a crucial role, especially since the aborted IPO of Ant Group in 2020, in shaping the sector. The People's Bank of China is the most important regulator in the space and formulator of policy, while the China Securities Regulatory Commission and the China Banking and Insurance Regulatory Commission play more operational supporting roles supervising elements of the fintech ecosystem.

The PBOC has also increasingly moved from a position exclusively as a regulator to more of a direct participant in the fintech ecosystem. This started with the creation of NetsUnion in 2017, a payment system to take over the movement of money between banks and payment companies and transactions that cross over between Alipay and WeChat Pay. It is now poised to reshape the landscape for digital payments and fintech through its development of a central bank digital currency (CBDC) which it calls the eCNY. Launching a central bank digital currency entails creating a payment system, which is likely to have both a collaborative and competitive role with fintech. Alipay and WeChat wallets can support eCNY transactions, but the eCNY is also designed to provide a separate 'backup' (Mu Changchun) system. Some design elements, like the central role of the largest banks, suggest a policy goal to boost the role of the state banks in the payment system, with greater use of their wallets instead of the private sector fintech firms.

## CHINESE FINTECH COMPARED TO OTHERS

Fintech in China dwarfs other countries. It has the largest number of users and transaction volume of mobile payments systems, the largest outstanding fintech credit, and the most valuable fintech firms in the world. It also the highest overall penetration of fintech as a portion of its population, at 87 percent, higher than the Netherlands' 73 percent and far larger than the United States' paltry 46 percent (EY 2019). The role of big tech firms was crucial, and here China is an outlier, with big tech facilitated payments at more than 16 percent of GDP, compared to the United States at only 0.6 percent (Figure 4.1).

In terms of central bank digital currency, almost all central banks are exploring issuance of this new type of currency, and a few smaller countries have already issued one. Among major economies, however, China is the furthest along, having committed to launch one back in 2016, back when top officials at the Federal Reserve in the United States had not even publicly mentioned the prospect (Chorzempa 2021). Concretely, its leadership in CBDC makes it a hub for knowledge on trade-offs in these systems and makes it able to be an early mover for any future cross-border infrastructure that transacts CBDC. China is involved in experimental proofs of concept for this infrastructure, as are central banks like the Bank of England, Bank of Canada, Bank of Japan, and the European Central Bank. Where it is different than those is on the retail side, where the others have not committed to launching one or have had very limited domestic pilots. In China around a quarter billion people have downloaded the eCNY wallets, giving the PBOC a unique set of practical knowledge around these systems.

Central banks tend to focus on operating payment systems between financial institutions, leaving many thorny issues of retail payments to the private sector. This makes it a steep but useful learning curve for the PBOC to be

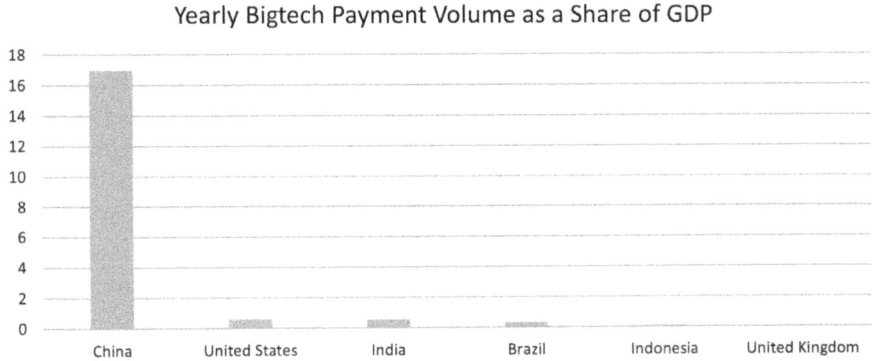


**Figure 4.2. Yearly Bigtech Payment Volume as a Share of GDP**
Source: BIS (2019).

on as it builds elements like developing its own app for the eCNY wallets, 'you're your customer' systems to check identities, onboarding procedures for millions of individual merchants that will accept its payment tools, algorithms that identify and stop fraud, and the regulatory system to manage cooperation and regulation with banks and mobile wallets that form part of the system. It will also see what cyber threats arise with the limited scale pilots, where glitches in code create issues, and the kind of payment volumes it can handle with different architecture.

## CHINESE GOALS AND PROGRESS TOWARD THEM

China's goals have shifted as fintech grew from a fledgling upstart to the way that most Chinese organise their financial lives. Initial encouragement and a hands-off attitude largely paid off as a thriving fintech sector seemed to replace a backward system, bringing it closer to achieving goals like financial inclusion, larger digitisation of China's economy, increased competition for traditional finance that improved services, and a better image for China abroad as an innovative country. Yet, in 2017 China Communist Party General Secretary Xi Jinping convened a Politburo study session with a focus on 'financial security,' which turned into a sustained campaign against financial risk that would sweep up fintech as well. Fintech's large scale and much of the low hanging fruit of digitisation already being picked led to much more regulation (Chorzempa 2018).

The financial regulatory side advanced, held back to some extent by the political power of fintech firms and their allies. Meanwhile a global movement to better protect privacy and restrain monopolistic practices among big tech firms changed policy goals. The Politburo's commitment on December 11, 2020 (Xinhua 2020), to 'prevent the disorderly expansion of capital' was precipitated by Alibaba founder Jack Ma's comments at a major conference that suggested that regulatory policy was excessively focused on reducing risk to the detriment of innovation and growth—just before what was to be a record breaking initial public offering of his fintech giant Ant Group. Pressure had already been building to rein in big tech, and Ma's comments appeared to be a public rebuke of Xi. Xi had personally advocated for a focus on financial security, and anyone willing to openly contradict this message could be considered a political threat. The political and economic issues were intertwined as well—if Ma could use public influence to shape regulatory policy in his interest, Ant Group could engage in riskier activities to increase profits. Much of ultimate costs of those risks, considering Ant could be considered too big to fail, would be borne by the public and the state.

Regulators then had to implement the Politburo's policy direction in fintech, identifying the areas in which order needed to be established and where capital had expanded too much. They had to tread carefully to avoid creating more disorder through excessively harsh policies that could disrupt now essential fintech services or excessively reduce economic growth. The fintech policies would fall under the overarching focus on the 'platform economy.' A March 2021 meeting of the Central Financial and Economic Committee, which Xi himself presided over, focused on the 'healthy and sustainable development of the platform economy' (Xinhua 2021). The meeting assessed that 'the overall development situation is good, and the impact is positive,' citing positive effects on efficiency and 'informatisation.' It signaled that regulatory authorities would need to ensure their anti-monopoly, financial risk reduction, and prevention of capital's disorderly expansion did not jeopardise the platform economy's benefits.

The first vision of officials' ideal fintech sector was in their plan for 'rectification' of Ant Group, still the most successful and prominent fintech firm in China, after authorities cancelled its IPO. PBOC Vice Governor Pan Gongsheng made clear that fintech was still encouraged to be more innovative and more competitive, just within regulatory constraints (PBOC 2021). Vice governor Pan's main policy goals revolved around financial risk, ensuring fintech remained competitive and inclusive, and improving both competition and data privacy.

## FINANCIAL RISK AND INCLUSION

Managing and reducing financial risk is the first and probably the most important of China's goals for fintech, and it is one where policymakers have achieved extraordinary success by reining in not only fintech, but also other types of loosely regulated so-called shadow banking. In 2018, the peer-to-peer lending sector, which at one point created around 5 percent of credit in China, was full of Ponzi schemes waiting to implode, with thousands of online platforms suffering from massive losses on poorly thought-out loans on the one side and owing millions of investors over a trillion RMB on the other. Authorities gradually diffused the risk and shut down the entire sector with minimal fallout to the financial system and social stability, the latter thanks to the quick mobilisation of the state security apparatus to discourage protests among investors taking losses in failed platforms (Chorzempa 2022).

In the payments space, risk from the rise of loosely regulated online payments has been diffused, as the state has ensured that all on and off ramps to online payment tools like Alipay and WeChat pay are regulated through a state clearinghouse called NetsUnion. All funds in these digital wallets are in turn deposited in special accounts at the central bank to ensure that payments companies cannot gamble with customer funds, and that if Alipay were to fail, customers would be easily made whole by the central bank. Security standards have tightened as well, ensuring that initially unsecure 'static' QR codes have moved to 'dynamic' ones that are much harder for criminals to use to drain people's digital wallets by sneaking a picture of the codes they present to pay.

Authorities have also successfully pressured Ant to slow down its consumer loans business and restructure it. The previous version allowed regulatory arbitrage, in which Ant's affiliated microloan company regulated at the provincial level originated hundreds of billions of RMB in loans nationwide that would then be sold to banks—meaning that Ant made money from the payments and loan origination while its IPO prospectus revealed that banks were on the hook for 98 percent of the credit risk (Ant 2020). Instead, it will have to issue loans through better regulated joint ventures.

Most importantly, regulators have created a financial holding company (FHC) regime to ensure that firms like Ant with multiple financial licenses along with non-financial business, are regulated at the group level. Previously, regulators had an incomplete picture because different pieces of Ant and other financial conglomerates were regulated by different authorities, with no overarching framework that takes into account the risks posed by interconnection of, say, payments with credit, investment, and banking. While this has not been a major problem in fintech thus far, other major financial instability has resulted when it turned out the banks like Baoshang Bank affiliated with commercial companies made risky, underpriced loans to their parent firms and related companies, leaving the government to foot the bill and make depositors whole.

Ant has been required to put its entire business in the financial holding company structure, but the process is not yet complete. The regime is a work in progress, and it will have to balance between risk considerations and ensuring it does not micromanage firms to the point of choking off useful product innovation, for example, making firms wait months for approval to launch a new product. However, its existence closes a major loophole in financial rules that fintech benefitted from for years.

Draconian bans on cryptocurrencies and exchanges used to trade them have been successful at stamping out a large share of this activity in China, which policymakers see as focused on risky speculation without benefit for the real economy. Though there may be a trade-off if cryptocurrency-based 'Web3' becomes a source of real innovation instead of mostly speculation and overhyped Ponzi schemes, but for now China is effectively avoiding the instability, losses, and gambling nature of cryptocurrencies, probably for the better.

On ensuring financial services are inclusive, Chinese officials assess that they have broadly achieved their goals, as 'basic financial services are now accessible in almost all urban and rural areas' (Liu 2021). This assessment is broadly corroborated, but work remains to be done. Fintech has made major contributions to achieving China's financial inclusion goals. Researchers exploring the spread and depth of use of digital financial tools found that these were predominately in use by residents of China's most prosperous regions in 2011, but by 2018 these were in widespread use across China, reducing inequality of access (Guo et al. 2019).

World Bank data show that from 2017 to 2021, the gap between digital payment use between the general population and the poorest 40 percent has halved. Nearly 80 percent of China's poorest citizens use digital payments (Demirgüç-Kunt et al 2022). Yet, despite this progress, China still has 130 million unbanked people, many of which will be difficult to reach due to illiteracy, lack of a mobile phone, or their remote location. There is still a long way to go to universal financial inclusion (World Bank 2022), but China's achievements put it not far from advanced economies, in which around 90 percent of adult populations use digital payments.

## RMB INTERNATIONALISATION

The international penetration of China's private fintech wallets is impressive, but still lags far beyond US financial giants in global reach. Alipay and WeChat pay are both accepted in dozens of countries for payments, so Chinese people can for example go to Thailand and buy things with Alipay. However, both Alipay and Tencent have made little progress towards internationalising their user base, e.g., gaining users abroad for Alipay and WeChat Pay that would make them more direct competition for multinational payments giants like Visa and Mastercard. Both Ant Group and Tencent have invested billions in fintech startups around the world, but this has not yet become a global network for them (Chorzempa 2022).

China's lead in fintech has done little for China's longstanding goal to internationalise the Renminbi, which punches far below China's economic weight in terms of its use in trade, investment, international reserves, and payments. The current dominance of the US dollar makes China vulnerable to US financial sanctions and changes in US financial conditions. For many years, the RMB has used for about 2 percent of transactions on the world's largest system for cross-border payments, making it the fifth most used payments currency after the US dollar, the pound, the euro, and the Japanese yen (SWIFT 2022). In terms of global official reserves the RMB's share has grown significantly from just more than 1 percent in 2017 to 2.88 percent, but it still ranked fifth (International Monetary Fund 2022). Meanwhile the US dollar made up nearly 60 percent. Overall, China has prioritised domestic control and stability, with strict capital controls, over a greater international role for its currency. Technological upgrades and China's growing role in trade and finance are not enough for greater Chinese leadership in this space.

China's reliance on the Society for Worldwide Interbank Financial Telecommunications (SWIFT), a communications system for payments, for cross-border payments is another point of vulnerability that China's fintech progress has made limited progress to removing. Though SWIFT is based in Belgium, the threat of 'secondary' sanctions (applying sanctions to entities that deals with sanctioned entities) against SWIFT have led it to disconnect Iranian and now many Russian banks under US sanctions from the system. China's own financial messaging system, Cross-Border Interbank Payment System (CIPS) is not yet a substitute for SWIFT and in fact incorporates SWIFT standards and messaging. It has only about one-tenth of the reach of SWIFT (Jin 2022). In April 2022, CIPS handled 14,500 transactions per day on average, 0.03 percent of SWIFT's average of more than 46 million per day in 2022.

Beijing hopes that the move to central bank digital currencies will provide a reset for international payments, potentially obviating the need for the US dollar or SWIFT. However, it has not yet made headway despite its leading status among major economies in central bank digital currency development. The PBOC's head of digital currency, Mu Changchun, proposed in 2021 a new foreign exchange trading platform to enable exchange over 'virtual borders' between digital wallets, synchronising elements of financial infrastructure to facilitate cross-border payments (Mu 2021), but there has not been a follow up to flesh out this idea. Cross-border CBDC pilots are happening among many central banks, including one at the Bank for International Settlements that includes the PBOC and a few other central banks, but these are in the very early proof of concept stage, Meanwhile the advanced trial work the PBOC has done has focused on rolling out a domestic currency and payment system that will compete and cooperate with the big private fintech payment offerings.

Overall, despite the shock of Russia sanctions creating more urgency for sanctions-proofing China's financial system, China still has a long road to go before tools like the eCNY, CIPS, or other financial infrastructures could make a real dent in its dependence on the US dollar and infrastructures it can influence.

## COMPETITION AND PRIVACY

The PBOC recognised that the Alipay/Tencent duopoly in online payments extended fintech giants' power into other markets, reducing competition. For example, with half a billion consumers using Alipay wallets to make purchases, Ant Group could provide an advantage to its own credit offerings by 'nesting' credit within its payment system, allowing users to buy seamlessly with credit at the point of sale (PBOC 2021b). The PBOC also had concerns that the easy availability of payment on credit, sometimes to the extent of making such systems the default means of payment, was encouraging excessive consumption and indebtedness (Guo 2020).

Authorities worried that network effects in payments could lead to an oligopolistic market structure in which 'winners take most.' It becomes a chicken and egg problem: merchants have little incentive to invest precious resources required to accept payment systems other than Alipay and WeChat pay because too few consumers use other payment systems, and few consumers have an incentive to use other payment systems because they are not accepted at merchants. The current state of the market for payments fits the PBOC's definition of monopoly, in which the top two players control more than two-thirds of the market (PBOC 2021a). The eCNY could raise competition once it is launched, but it could also become a stultifying force against innovation, a state takeover of retail payments, considering that state-backed retail payments projects like UnionPay have a poor track record for innovation. Authorities have a vision for competition to improve involving data portability and ownership rights of individuals, breaking up "data islands" of big tech firms to allow their data to be used by other firms, and reduction of the advantages the big platforms have through more separation between the financial and nonfinancial operations (Yang and Potkin 2022). However, these ideas will be technically difficult to achieve, and some may come at a trade-off to other policy goals, for example, increased data sharing with privacy. They will take years to implement.

On privacy/data protection the rules already tightened substantially in the late 2010s, for example when Ant group was forced to change its data sharing practices within the group for credit scoring. Privacy is an area with strong trade-offs, for example, limited data sharing to protect privacy means that one's credit history with one provider cannot be shared with another. This limits competition because only one lender can effectively evaluate the borrower's creditworthiness, and it makes lending riskier because lenders cannot know if a potential borrower has already failed to pay back loans from another creditor. The current vision is to force Ant Group to share its data, not just credit histories but other more sensitive data, with a joint venture together with state firms, which would then help lenders, including the state-owned banks, leverage these data to provide credit. This is supposed to make that firm's outputs more objective and reduce conflicts of interest through independence from Ant, which competes as a lender with the institutions it supplies with data or credit scores.

## CONCLUSION

Chinese authorities have achieved many of their long-term goals for the fintech sector, but balancing competing and shifting objectives will be a major future challenge. Financial technology innovation has which has helped Chinese internet firms grow, becoming a crucial enabler for the development of China's digital economy. Hundreds of millions of Chinese now have abundant choice for payments, investments, and loans, and together with other services in super apps they have transformed daily life in China. Still, the challenge remains that China's financial system remains dominated, even if not as much as before, by state-owned banks with often non-commercial incentives, outdated IT systems, and mindsets around credit that struggle to effectively adopt new technology that can make things more efficient. A continued larger role for state companies, which are easier to control, does not bode well for the future of fintech innovation, which has almost exclusively been driven by private firms.

Authorities' goal is to marry the fintech innovators and their data with the capital advantages and political reliability of the state banks, but this has not yet been achieved. Instead, they have restrained some of the consumer credit especially of the big fintech firms, which has contributed (though swamped by the zero covid disruptions) to the slowdown in consumption. Now, authorities have to balance these traditional financial trade-offs with those in a complex new set of issues.

One challenge is balancing privacy and competition. Ideas for opening up the data troves of the big tech companies to other firms will make it more difficult to ensure these data are protected. For example, hackers could use fake user requests to download "their" data or exploit the systems under consideration to allow big tech firm data to flow to startups. Another is sustaining innovation in fintech, which may struggle under a much higher regulatory burden. More approvals will be needed, which will slow down the process, and authorities have become more risk averse. Efforts to internationalize the RMB and have globally competitive fintech firms are more likely to face barriers abroad if they are perceived internationally as operating as tools of the state. It is harder to justify allowing them to collect sensitive data abroad due to the state's assertion of increased control domestically, some cases forcing fintech firms into joint ventures with state firms, and integration into the surveillance apparatus, like their hosting of Covid apps with tracking functions.

## REFERENCES

Analysys. 2022. Yidong Zhifu Hangye Shuzihua Jincheng Fenxi: (Mobile payment industry digitization analysis). 30 September. Available from: https://www.analysys.cn/article/detail/20019936 [5 July 2022].

Ant Group. 2020. "IPO Prospectus" Available from: https://web.archive.org/web/20201105212227/https://www1.hkexnews.hk/app/sehk/2020/102484/documents/sehk20082500535.pdf [5 July 2022].

Bank for International Settlements. 2019. Annual Economic Report. Available from: https://www.bis.org/publ/arpdf/ar2019e3.htm [12 January 2022].

Barnett, W and Lorentzen, P. 2006. 'EachNet.com,' Stanford Business School Case No. SM91. Available from: www.gsb.stanford.edu/faculty-research/case-studies/eachnetcom [16 November 2022].

Chorzempa, M. 2018. 'Beijing's Grip on Internet Finance Is Tightening' 9 January. 2018. *China Economic Watch.* Peterson Institute for International Economics. https://www.piie.com/blogs/china-economic-watch/beijings-grip-internet-finance-tightening [16 November 2022].

Chorzempa, M. 2021. 'China, the United States, and Central Bank Digital Currencies: How Important Is It to Be First?' *China Economic Journal* 14(1), 102–15.

Chorzempa, M. 2022. *The Cashless Revolution: China's Reinvention of Money and the End of America's Domination of Finance and Technology*, New York: Public Affairs.

Cornelli G., Frost J., Gambacorta L., Rao R., and Ziegler T. 2020. Fintech and big tech credit: A new database. *BIS Working Papers* no. 887. Bank for International Settlements. Available from: https://www.bis.org/publ/work887.pdf [13 January 2022].

Demirgüç-Kunt, A., Leora K., Singer, D. and Ansar, S. 2022. Global Findex Database 2021: Financial Inclusion, Digital Payments, and Resilience in the Age of COVID-19. Washington, DC: World Bank. doi:10.1596/978-1-4648-1897-4.

Engen, J. 2018. 'Lessons from a mobile payments revolution,' *American Banker*. Available from: https://www.americanbanker.com/news/why-chinas-mobile-payments-revolution-matters-for-us-bankers [16 November 2022].

EY. 2019 Global Fintech Adoption Index. Available from: https://www.ey.com/en_gl/banking-capital-markets/what-is-next-for-asia-in-fintech-adoption [16 November 2022].

Guo Feng, Wang Jingyi, Wang Fang, Kong Tao, Zhang Xun, Cheng Zhiyun. 2019. 'Measuring China's Digital Financial Inclusion: Index Compilation and Spatial Characteristics,' Working paper. Institute of Digital Finance. Peking University. Available from: https://en.idf.pku.edu.cn/docs/20190610145822397835.pdf.

Guo, Shuqing. 2020. 'FinTech developments, challenges and supervision in China.' Video Speech at the Singapore FinTech Festival (SFF) 2020. 8 December.

Available from: https://www.caixinglobal.com/upload/guo-shu-qing.pdf [18 November 2022].

International Monetary Fund. 2022. 'Currency Composition of Official Foreign Exchange Reserves (COFER).' 20 June. Available from: https://data.imf.org/?sk=E6A5F467-C14B-4AA8-9F6D-5A09EC4E62A4 [4 July 2022].

Jin. E. 2022 'Why China's CIPS Matters (and Not for the Reasons You Think).' 5 April. *Lawfare.* Available from: https://www.lawfareblog.com/why-chinas-cips-matters-and-not-reasons-you-think [4 July 2022].

Liu, Guiping. 2021. On the Development of Financial Inclusion in China. *China Finance* Issue 16. December. Available from: http://www.pbc.gov.cn/en/resource/cms/2021/10/2021100910485035020.pdf [4 July 2022].

Ma, J. 2018. Speech at World Economic Forum. 24 January. Available from: https://www.youtube.com/watch?v=4zzVjonyHcQ [4 July 2022].

McGregor, R. 2012. '*The Party: The Secret World of China's Communist Rulers.*' New York: Harper Perennial.

Mu, Changchun. 2021. 'Speech at the BIS Innovation Summit.' 25 March. Available from: https://www.bis.org/events/bis_innovation_summit_2021/agenda.htm?25_March_2021=2 [4 July 2022].

People's Bank of China. 2005. 'A PBC Official Answers Questions of Reporters on the E-payment Guidance, no. 1.' 26 October. Available from: https://web.archive.org/web/20051230034447/http://www.pbc.gov.cn/english/detail.asp?col=6400&id=611 [4 July 2022].

People's Bank of China. 2021. 'Zhongguo Renmin Yinhang Guanyu Fei Yinhang Zhifu Jigou Tiao Li (Zhengqiu Yijian Ga) Gongkai Zhengqiu Yijian de Tongzhi' (Notice from the People's Bank of China concerning opening the Nonbank Payment Institution Regulations (Draft for Public Comments) for public comment). 20 January. Available from: http://www.pbc.gov.cn/rmyh/105208/4166553/index.html [4 July 2022].

People's Bank of China. 2021. 'Zhongguo Renmin Yinhang Fuhangzhang Pan Gongsheng Jiu Jinrong Guanli Bumen Zaici Yuetan Mayi Jiyuan Qingkuang Da Jizhe Wen' (People's Bank of China Vice Governor Pan Gongsheng Answers Questions on Financial Regulatory Authorities Once Again Calling Ant Group to Discuss its Situation). 中国人民银行副行长潘功胜就金融管理部门再次约谈蚂蚁集团情况答记者问. 12 April. Available from: http://www.pbc.gov.cn/goutongjiaoliu/113456/113469/4229432/index.html [4 July 2022].

Reuters. 2012. 'China's Wen Urges Breakup of Bank Monopoly as Growth Slows,' *Reuters*. 3 April. Available from: www.reuters.com/article/us-china-banks/chinas-wen-urges-breakup-of-bank-monopoly-as-growth-slows-idUSBRE83211C20120404 [18 November 2022].

Yang, Yingzhi and Potkin, F. 2022. 'Ant, Alibaba plan for less intertwined future after China crackdown,' *Reuters.* 21 June. Available from: https://www.reuters.com/world/china/ant-alibaba-plan-less-intertwined-future-after-china-crackdown-2022-06-21 [18 November 2022].

Society for Worldwide Interbank Financial Telecommunication. 2022. 'RMB Tracker.' June. Available at: https://www.swift.com/our-solutions/compliance

-and-shared-services/business-intelligence/renminbi/rmb-tracker/rmb-tracker-document-centre [18 November 2022].

State Council of the People's Republic of China. 2005. 'Guowuyuan Bangongting Guanyu Jiakuai Dianzi Shangwu Fazhan de Ruogan Yijian' (State Council General Office opinions on how to accelerate e-commerce's development). Issued January 8. Available from: www.gov.cn/gongbao/content/2005/content_63341.htm [18 November 2022].

Xiaobo Wu. 2017. *Tengxun Zhuan 1998–2016: Zhongguo Hulianwang Gongsi Jinhua Lun* (Tencent biography 1998–2016: Theory of evolution of China's internet company] Hangzhou: Zhejiang University Press, 39.

Xinhua News. 2021. 'Xi Jinping Zhuchi Zhaokai Zhongyang Caijing Weiyuanhui Di Jiu Ci Hui Yi' (Xi Jinping Presided over the Ninth Meeting of the Central Financial and Economic Work Commission).' 15 March. Available from: http://www.gov.cn/xinwen/2021-03/15/content_5593154.htm [18 November 2022].

Xinhua News. 2020. 'Zhonggong Zhongyang Zhengzhi Ju Zhaokai Huiyi, Xi Jinping Chuxi' (Xi Jinping chairs a meeting of the Political Bureau of the Central Committee of the Chinese Communist Party).' 11 December. Available from: http://www.xinhuanet.com/politics/leaders/2020-12/11/c_1126850644.htm [18 November 2022].

# PART III

# International Engagement and Confrontation

# *Chapter 5*

# China

## *A Technical Standardisation Power?*

Tim Rühlig

When the Central Committee of the Chinese Communist Party (CCPCC) and the State Council of the People's Republic of China (PRC) jointly published China's new "Standardisation Outline" (in the following simple the "Outline") in October 2021, the crucial importance of technical standard-setting contrasted with a remarkably modest description of China's aspirations. As the quote above illustrates, the Chinese leaders describe technical standards having a "fundamental and leading role" for national governance and acknowledge an "urgent need" to strengthen technical standardisation to "build a modern socialist country." These are high expectations for voluntary and highly technical documents that are mostly developed by engineers from companies and research institutions.

While the Outline formulates a comprehensive reform plan that would strengthen China's technical standardisation capabilities, it falls far short of a "grand strategy" to global domination in this field as some might have expected. Referring to "China Standards 2035," an often-misunderstood attempt of some actors within the Chinese party-state to push for further reform, technical standardisation is more and more perceived through the lenses of geopolitical competition. However, "China Standards 2035" was a research project concluded launched by reform-oriented elites from within the party-state. The results of the project never became public. While the Outline contains some of its results it also deviates from "China Standards 2035." This is illustrative of the fact that China's standardisation approach remains contested within the party-state.

Considering the emerging competition over high technology between the United States and the PRC, observers on both sides of the Pacific have identified technical standardisation as an arena of this growing rivalry. For example, US Secretary of State Antony Blinken advocates closer cooperation with allies and partners on international technological standard-setting as a means to preserve and expand US strategic influence (Reuters 2020). China's ambitions are not only formulated in the Outline. For instance, domestic and international standard-setting targets are referenced in a wide range of sections and chapters of the 14th Five-Year Plan (Xinhua 2021). Politicians may need to constrain themselves. But academics, journalists, and think tanks in the PRC, the United States, and the European Union (EU) describe technical standardisation as an arena of the power competition over high technology (Pop, Hua and Michaels 2021; Atlantic Council 2021; Yan 2020; Seaman 2020, 21–2). Their line of argument is based on three assumptions that might be reasonable but are hardly ever examined, although they are anything but a given.

First, US, Chinese, and European observers seem to assume that the ability to shape technical standards comes with power advantages to the respective states and broader influence in international affairs. To those familiar with technical standard-setting this might not sound entirely unfamiliar, but it is also counterintuitive. For this, one needs to properly understand what technical standards are: Technical standards are omnipresent shaping our lives, but they are *voluntary* product specifications that generate basic safety and interoperability. They are either the result of market dominance (*de facto* standards) or developed by standard developing organisations (SDOs) that overwhelmingly consist of representatives of industry. While public actors do play a (minor) role, technical standardisation is largely an example of private self-regulation. In contrast patents, a good standard is broadly available and accepted globally (Deron 2020). If a technical standard is released but not used by the market it is ineffective. Hence, technical standards require market acceptance and need to be available to all market participants. Where technical standards consist of patented technologies, we speak of standard-essential patents (SEPs), patent holders are obliged to license their patents under fair, reasonable, and non-discriminatory terms (FRAND). Courts around the globe are enforcing FRAND terms on patent holders.1 Hence, technical standards apply a very different logic than, for example, export controls or punitive tariffs that aim to exclude competitors from supply or hinder market access. While these means are exclusive, technical standards gain their effectiveness from their inclusivity and availability. In the emerging technological competition, the United States and China primarily use such measures. This raises the question why technical standards should be available to the exclusionary logic that underlies the emerging technology competition.

Similarly, the political impact is also not obvious from the technical nature of these standards. USB is a standard for cables, connectors, and protocols that enable charging and the exchange of data on a wide range of devices. Similarly, Wi-Fi is a family of radio technologies built on technical standards that allow for wireless local area networking (WLAN) of a wide range of technological equipment. Technical standards allow products of all kinds to be applicable in a multitude of contexts across countries and manufacturers. Hence, technical standards are highly specific technological product specifications, developed by private industry and accessible to everyone. It is not obvious how the host state of a company that has suggested a given technical solution to become a standard politically profit from the proposal's success.

Second, the debate on technical standards as part of the emerging competition over high technology mostly takes for granted that China possesses increasing impact on standard-setting. In the political debate more specifically, many observers assume that China is about to "dominate" international technical standard-setting (Gargeyas 2021). This bears the question how to precisely measure technical standardisation power and whether China's footprint2 is factually dominant or reaching a state of domination.

Third, in the West, China's growing "standardisation power" is mostly discussed in an alarming tone (SFRC Democratic Staff 2020). Not least historical analogies are introduced to argue that standard-setting can have devastating impact if great powers ignore growing engagement by rising powers (Brookings Institution 2021). While such line of argument might not be unreasonable it mostly lacks a clear concept of "standardisation power." This absence hinders observers to be precise about the inherent risks.

This chapter addresses three questions to better understand the importance, the ambitions, the achievements, and the risks of China's technical standardisation approach. First, this chapter asks why technical standardisation power is important for China. Second, the chapter explores to what extent China has already turned into a technical standardisation power. Third and finally, the chapter assesses how different China's policy is from that of the rest of the world and what the potential inherent risks of China's growing presence in technical standardisation are.

## WHY IS STANDARDISATION POWER IMPORTANT FOR CHINA?

The ability to shape technical standards bears enormous influence that is crucial for the PRC—both domestically and internationally. This section substantiates this claim developing a four-dimensional heuristic of standardisation power. It differentiates between economic, legal, political, and ideational dimensions.

*Economic dimension:* Technical standards have distributary effects. These are the results of a high share of technical standards consisting of patented technology. In the field of Information and Communication Technology (ICT) standards, an estimated 55 percent is patented.3 As briefly mentioned at the beginning of this chapter, patents and technical standards have contrary purposes. Technical standards follow an inclusive logic because they help spread technological solutions across manufactures to establish interoperability and guarantee basic safety. In contrast, patents protect inventions and exclude competitors from using them. When technical standards consist of patented technology, this contradiction is resolved through licensing. When patent holders declare SEPs, they commit to licensing on FRAND terms. FRAND might sound like there is no competition involved, but this impression is false. SEPs are available to all in return for potentially enormous royalty fees that manufacturers must pay to SEPs holders. The Swedish technology company, Ericsson, for example, earned €5.2 billion by licensing technology in 2017, accounting for more than 20 percent of the company's revenue (Strumpf 2019). Chinese technology firms such as Huawei do not provide precise information on the share of royalties to their revenue. However, field research confirms that licensing plays a crucial role to several Chinese technology giants and generates alternative revenue streams over manufacturing. In fact, in a situation when Chinese companies are excluded from the deployment of technology in any given third country as has been the case with the exclusion of Huawei from some Western markets, companies still profit by means of royalties. This can be a lifeline for companies that find themselves under pressure from sanctions regimes.

The distributary effects of technical standards are not limited to the payment of royalties for SEPs. Often, companies that fail to establish their technological solutions as technical standards need to redesign their products to comply with standards and thereby generate interoperability. Hence, those successfully setting international technical standards cannot only expect royalties but also avoid adaption costs. Given the considerable size of these distributary effects technical standards affect competitiveness.

Successfully setting technical standards of key-enabling technologies shapes national economic competitiveness. A competitive industry in strategic sectors can only be beneficial to the host state. In China, more specifically, party-state influence over strategic sectors makes it even easier to leverage industry strength for political purposes (Wu 2016; Williams 2018).

*Legal dimension:* De jure, international technical standards are voluntary technical specifications. Nonetheless, standards can unfold enormous legal force. The Agreement on Technical Barriers to Trade (TBT), the Agreement on Government Procurement (GPA), the review of the Agreement on Sanitary and Phytosanitary Measures (SPS) and the General Agreement on Trade in Services (GATS) under the framework of the World Trade Organisation (WTO) all treat international standards as benchmarks for the facilitation of international trade. Standards serve as important qualifications of what accounts as a legitimate exception, for example, under the pretext of basic safety requirements (Graz 2019, 89). Article VI: 5b GATS, for instance, stipulates that international standards of relevant international organisations serve as yardsticks to ensure that trade in services is not more burdensome than necessary (WTO 2012, 185–6). If domestic technical standards deviate from international ones, in principle, the judiciary of the WTO could find a state to be noncompliant with international trade law unless the respondent can provide a reasonable explanation for such deviation. This could be specific national circumstances necessary to protect human health and safety or the environment. Crucially, roughly 80 percent of international trade is affected by technical standards and associated technical regulations (OECD 1999). Hence, they hold a rather broad force.

Furthermore, domestic technical standards can have extraterritorial effects. States reference technical standards in legally binding documents, mostly in regulations. When regulations prescribe thresholds, technical standards can serve as a suggested method for upholding the limits set in a regulation. In other words, if a referenced standard is applied the respective product is assumed to be in conformity with the regulation. Companies seeking market access do not need to comply with the technical standard. Often, however, it is the easiest and cheapest option to implement the technical standard. When standards are referenced in the regulations of major markets such as the European Single Market, the United States, or the PRC, they can unfold extraterritorial effects because multinational companies often choose to comply with the strictest technical standard. This generates conformity with regulations and thereby market access to all relevant markets.

*Political dimension:* Global technical standards facilitate borderless globalisation. If contradictory technical standards exist in different geographical locations, however, they generate distinct technological spaces. Hence, depending on the scope of their validity, technical standards can either facilitate global trade or create geographically bifurcated technological corridors. The result of the latter can be lock-in effects with political costs beyond the field of technical standardisation itself.

For example, to date, technical standards in the railway sector remain largely national or regional. If country A adopted the national railway standards of country B—ranging from track gauges to traction technical parameters and voltage—the maintenance and further buildout of the railway cannot be carried out by suppliers other than those based in country B. Other suppliers produce noncompatible technology based on deviant technical standards.

Country A is locked into country B's technology and fully reliant on supply from country B. Economists have long referred to these lock-in effects as "network effects" or "network externalities" (Bonardi and Durand 2002). These studies have convincingly demonstrated that high switching costs lead to the preservation of dominant technical standards (Arthur 1989; Farrell and Saloner 1985); standards developed in an early stage often prevail, even if they are technologically inferior (Schilling 2002).

Politically, this remains unproblematic as long as the respective product is not of strategic importance to the well-being of a society. Railways, however, are a critical infrastructure enabling the flow of goods and people, thereby generating welfare and mobility. A lock-in effect in such a critical sector has political implications. If all suppliers that are compliant with the respective technical standards are based in country B, it could ask country A for political concessions in return for the maintenance and buildout of the critical infrastructure. Even if country B does not explicitly ask for such concessions, country A would think twice before adopting a confrontational stance on issues of core interest to country B in fear of the consequences for the functioning of its critical infrastructure.

Apart from such lock-in effects, some experts suggest that technical standards have the potential to impinge on what is often regarded as the crown jewel of state power: security. Regarding cybersecurity, those who develop a standard could have a deeper knowledge of the technology including its vulnerabilities. If adopted as an international standard, the technology spreads globally. Consequently, the developer of the technology in question possesses prime knowledge of its flaws that can be used to undermine an adversary's cybersecurity (Eisenstark 2018; Medin and Louie 2019). Other observers counter that standardisation is a process of maximum transparency in which it is hardly possible to hide security-relevant flaws from the eyes of the engineers of potential adversaries. From this perspective, a high degree of standardisation even increases the cybersecurity by means of transparency. Whichever perspective is more accurate, technical standardisation influences the degree of cybersecurity (Rühlig 2019).4

*Ideational dimension:* How technology is designed is highly political as it inscribes ethical values to it. Technology does not exist in a vacuum. Technical standards shape the physical world and contribute to the constitution of our social lives (Busch 2011). Hence, technical standards determine what is perceived as "normal" technology. Several critical scholars have described technical standards as social institutions in their own right (Krislov 1997; Hallström 2004; Timmermans and Epstein 2010). For instance, Wi-Fi is seldom questioned as the dominant WLAN standard. However, a few years after Wi-Fi had been established as the international standard, China proposed the WAPI technology as a new standard. WAPI promised better performance but provided worse privacy (Lee and Oh 2006). Finally, the Chinese proposal failed mostly due to procedural issues (Suttmeier, Yao, and Tan 2009). Whether intended or not, by rejecting WAPI international SDOs prioritised privacy over performance, shaping what consumers expect from WLAN technology.

This is not an isolated example. Emerging technologies are increasingly penetrating all spheres of public and private life. Whoever sets technical standards on algorithmic bias, data privacy and similar issues shapes ethical, political, and security angles of key enabling technologies (Seaman 2020). Moreover, the ideational power of technical standardisation is not limited to underlying ethical values. If a country shapes international technical standardisation, it is likely to gain a reputation as a technology leader, which is a sigh of societal progress beyond economic and military prowess.

## HAS CHINA ALREADY BECOME A STANDARDISATION POWER?

Measuring whether China is a standardisation power is more complex than one might think not least because technical standardisation spans a wide range of products and technologies and is in itself a highly technical process of negotiations among specialised engineers in which one proposal seldom fully prevails. Moreover, technical standards are developed in a multitude of international institutions. For example, more than two hundred international SDOs exist in ICT (Schneiderman 2015). To grasp China's influence, a total of six parameters need to be taken into account.

The first proxy is *leadership positions in international SDOs*. Almost all SDOs have an institutional leadership with varying degrees of influence. Since the nature, processes and composition of a great number and diversity of SDOs is varying it might not be easy to identify the most relevant SDOs. For a study that aims to grasp China's general influence on technical standardisation, the broadest and most famous international SDOs should be considered. This includes the International Standardization Organisation (ISO), in which China has become a permanent member of the institution's main governing bodies, the ISO Council (in 2008) and the ISO Technical Management Board (in 2013). From 2015 to 2018, Zhao Xiaogang was the first Chinese citizen serving as rotating ISO President. Similar in importance to ISO is the International Electrotechnical Commission (IEC) which is led by Zhu Yinbiao after having served as the IEC's Vice President. The third main international SDO, the International Telecommunication Union (ITU) is currently led by a Chinese official, Zhao Houlin. Before his election, he served as ITU's deputy Secretary-General.

These institutional leadership positions help shaping the agenda and are the result of preexisting power, but they come with relatively little impact on the actual standard-setting. For this process, technical leadership positions, often called "secretariats," are more crucial. Secretariats are supposed to be neutral (ISO 2018), but technical standardisation experts agree in that secretariats have an enormous influence by structuring, organising, and coordinating the process.[5] In ISO, China currently holds sixty-eight secretariats; five countries lead more technical committees. In IEC, China ranks seventh with a total of twelve secretariats. This illustrates that China is far from dominating international SDOs in terms of technical leadership positions. However, the proportion of China's influence is growing. In the period 2011–2018, China's share in ISO Technical Committee and Subcommittee secretariats grew from 5 percent to 8.21 percent, that of ISO Working Group secretariats even from 2 percent to 6.58 percent.[6]

In the ITU, China has achieved an even greater influence. Together with Japan, it holds the most ITU-T study group chair positions and is the sole leader in ITU-T study group vice-chair positions. The PRC is also the strongest of all nations in terms of ITU-T work program chair and vice-chair positions; the same holds for ITU-T rapporteur posts. In ITU-T Focus groups, China ranks second behind the United States in chairs, but it outnumbers all with regard to vice-chairs. In the Third Generation Partnership Project (3GPP), another important SDO in the telecommunications sector, China holds the most Working Group chairs and vice-chairs (DiploFoundation 2021).

Technical leadership positions are important but not a necessary requirement to impact standardisation. A second proxy, *participation in standard-developing committees*, captures which actors can submit proposals and comments to the standardisation process. China's influence has grown enormously since 2007 having surpassed that of the United States, France, and Japan. However, China still falls slightly short of the United Kingdom and Germany.

Another measure for participation is the number of participations. Data from 3GPP shows that, in 2018, China accounted for the highest share of participants (23.7 percent). Representatives from firms based in the EU and the United States fell slightly short with 22.5 percent each.[7] This corresponds with the fact that no other country has more individual members in 3GPP. ITU member statistics similarly indicate that among all countries, China falls only short of the United States.[8]

In many SDOs (including ISO and IEC), membership in standard-developing committee requires regular contributions. Otherwise, national SDOs lose the status as active participant. However, membership does not reveal the *number of contributions and whether they are adopted*. Hence, such data is considered to be the third proxy. Statistics on standard contributions are rare and necessarily incomplete, because technical standardisation encompasses a wide variety of different products and technologies. This chapter exemplarily draws on technical standardisation contributions to the development of one key-enabling technology that has received a lot of attention recently: 5G. Statistics demonstrate that China's share in 5G contributions ranks first with 31.5 percent and has increased from the previous generation of mobile technology, 4G/LTE, when Chinese companies accounted for around 22.4 percent of the contributions (data quoted in Pop, Hua, and Michaels 2021). When 5G standardisation contributions are compared to their adoption, however, data reveals that China does slightly worse than firms based in Europe (Pohlmann, Blind, and Heß 2020).

The data from 5G standardisation is in line with findings on China's strength in standardisation in mobile network technology more broadly. According to both measures, Europe is slightly ahead of the PRC.9 In the Internet Engineering Task Force (IETF), document contributions from the United States dominate with a share of more than 44 percent. China ranks second with 12.9 percent (DiploFoundation 2021).

In the public discussion of standardisation, another proxy widely used to capture the influence of an actor on standard-setting is *SEPs declarations*. This is the case because standards of many key-enabling technologies such as 5G consist to a great extent of patented technologies. Very often, however, the way SEPs are discussed is misleading. Existing data only captures declarations of standard-essentiality, which should not be confused with actual standard-essentiality. Measuring SEPs is difficult because the actual standard essentiality of many patents remains unclear. When a standardisation process starts, participating actors declare patents as standard-essential thereby indicating that they believe the respective patent could become essential for the standard and that they are willing to license the patent under FRAND terms. When a standard is established, no comprehensive analysis takes place that could establish whether the standard-essentiality declaration turned out to be correct. In some cases, firms file complaints against deviating assessments of standard essentiality by their competitors when demanding licensing fees. In most cases, however, technology companies negotiate package deals exchanging licensing fees for groups of patents without assessing the standard essentiality of each individual standard. While observers assume that all companies declare more of their patents as standard-essential than turns out to be correct, this holds true for all actors involved. In terms of 5G SEPs declarations, China is ahead of other countries with a share of 33 percent followed by South Korea with 27 percent (Pohlmann, Blind, and Heß 2020).

However, not every patent is technologically equal in importance to a given technology. Technological relevance is often calculated by means of the average size of a patent family and the average number of citations of the respective patent in other declared SEPs. While the size of the patent family is supposed to measure how extensive the patents are, the number of citations serves to indicate how relevant a certain patent is for other components of 5G technology by being referenced in other 5G-relevant patents. Based on the IPlytics database, Chinese patents turn out to be the least important compared with the those filed by companies from other major technological leaders in 5G based in Europe, the United States, South Korea, Japan, Taiwan, and Canada (ibid.).

While all these quantitative proxies consistently point to a growth in Chinese impact on international technical standard-setting, hardly any figure can capture the entire development. Therefore, a fifth proxy deals with the *qualitative description* of influence in the standardisation. Interviews with participants in international SDOs helps. For this chapter, more than seventy-five interviews with European and US representatives in international SDOs have been conducted. The results confirm the quantitative findings.10

> *China understands the value of standardisation. Our Chinese colleagues have become very active participating in large groups in meetings. At first, they had to learn the rules of the game. Over the last decade, they developed into an integral and influential part of standardisation community.* 11

While international participants in SDOs continue to see quality issues preventing a stronger Chinese influence, they also consistently acknowledge that China improves.

> *A decade ago, our Chinese colleagues could contribute good proposals in only very few fields. To this day, many Chinese proposals are rejected because their technological quality is inferior to the contributions of other experts. Regardless of these continuous challenges, you cannot ignore the improvements. The time when some of my colleagues would not take Chinese contributions seriously are gone.* 12

This development has sparked fears among standardisation experts from the United States and Europe that China could outstrip Western countries in international technical standardisation. One expert involved in ICT standardisation exemplarily warned:

> *if China's influence in ICT standardisation will continue to grow at the same pace, we will soon be on the receiving end. We are at a critical junction and are only slowly waking up. [. . .] We need to get better and live up to the challenge.* 13

Chinese standardisation experts confirm in more than thirty interviews that their work has turned into a priority within China and that they believe they have made significant progress.14

> *Five or ten years ago, we could not make a lot of real contributions to international technical standardisation. Our own innovation was not good enough, but we also did not quite understand the importance of standard-setting. [. . .] We are proud to say we have made much progress. But we still need to learn a great deal from Europe and the US.*15

Finally, technical standards are not only developed in SDOs, but can also be established as de facto standards. The most prominent examples are the operating systems of Microsoft and Apple that have never been adopted by an SDO. However, the fact that any software needs to be compatible with Windows or iOS if it does not want to end up in a niche makes both operating systems de facto standards. Quantitative accounts of de facto standards are hardly feasible, which requires qualitative investigations including interviews. If one aims to generally grasp impact in de facto standard-setting, a focus on specific vehicles can be helpful. As an example, this chapter explores the *role of the BRI* as a sixth proxy for China's influence in international standardisation.

To begin with, China's BRI includes an explicit standardisation dimension. In 2015, China's main macroeconomic agency, the National Development and Reform Commission (NDRC), issued its first "Action Plan for the Harmonisation of Standards along the Belt and Road" (PCR 2015). In late-2017, the NDRC published another action plan setting further benchmarks.16 As part of the plan, China began to translate its domestic technical standards into foreign languages to facilitate their adoption in third countries. By September 2019, China had signed ninety bilateral agreements on technical standardisation cooperation with fifty-two countries and regions.17 Chinese experts acknowledge, however, that the agreements are vague and often meaningless. A major state-sponsored research project, "China Standards 2035," suggests transforming these agreements into a regional technical standardisation organisation, the BRI Standards Forum, that could develop BRI Regional Standards.18 Whether such a Forum could fulfill the ambitious goal of developing regional standards that are acknowledged along the BRI remains to be seen. The Forum, if established, could also simply serve to coordinate activities in ISO and IEC with the potential to further strengthen Chinese influence in these institutions.

More importantly, many concrete BRI projects incorporate Chinese technical standards. It is through these projects that the PRC disseminates its domestic technical standards to third countries without submitting them to international SDOs, as in the case of railway standards mentioned above. In sum, all proxies indicate a growing footprint though to different degrees. In other SDOs such as the Internet Engineering Task Force (IETF) or the Internet Corporation for Assigned Names and Numbers (ICANN), China's influence remains scarce. Hence, China does not dominate international technical standardisation, but its ability to shape standards grows steadily.

## HOW UNIQUE IS CHINA'S STANDARDISATION APPROACH?

As I have shown elsewhere in detail, China's approach to technical standardisation is state-centric and deviates significantly from wester systems (Rühlig 2020). Until 2018, China's domestic standardisation was formally entirely public. All three types of technical standards—national, local, and sectoral—were developed under the auspices of ministries or local governments. With the growing importance of private companies, private firms were increasingly involved in standard setting, but always within the institutional framework of state ministries and local governments. A significant share of what China referred to as standards was mandatory.

The new Standardisation Law that came into force on 1 January 2018 institutionalised the increasing role of the private sector in Chinese standardisation (SAC 2017). Technical standards are now developed in two tiers, one state-driven and one market-driven. National, local, and sectoral standards continue to exist, representing the state tier. All local and almost all sectoral standards are voluntary, and the number of mandatory national standards was massively reduced. While this reform was clearly inspired by European and US standardisation practices, it remains a unique, state-centric approach (Rühlig and ten Brink 2021).

Even in the market-tier many SDOs have very close ties to the party-state. One way for the authorities to continuously influence market-driven standard setting is through informal guidance. Party-state turns out to be essential for SDOs to become influential in China. Since 2018, more than 2,700 industry federations have registered almost 10,000 association standards, many of them conflicting with one another. Interviews with Chinese firms and international firms operating in China indicate that different forms of party-state support are decisive for whether a given SDO is more relevant than its competitors in China.

As I have demonstrated elsewhere in detail, China externalises its state-centric standardisation in both formal and de facto standard setting to the international arena. Consider the case of 5G standardisation. Strikingly, China's formal 5G standardisation practices are not markedly different from those of the West. However, the PRC has adapted these practices to its state-centric approach thereby externalising its domestic standardisation. For example, party-state investment and guidance has been crucial to increase Chinese companies' technical expertise. Similarly, state funding has been made available for active participation in international standardisation bodies.

In order to exploit first-mover advantage, a central feature of the party-state's industrial policy is to establish regulatory and financial conditions to facilitate early commercialisation of key enabling technologies. In 5G, the PRC has not only sponsored the world's largest 5G trial area in the Yangtse River Delta, but the state-controlled mobile operators have been instructed to roll out the most innovative version of 5G, known as standalone 5G (Shi-Kupfer and Ohlberg 2019). Western countries, in contrast, have tended to opt for the less innovative update of 4G/LTE networks to non-standalone 5G because private industry has identified that this path requires less investment and is therefore more economical in the short and medium term (Eisenstark 2018; Rühlig and Björk 2020).

Another example of party-state involvement is the coordination in order to speak with one voice. Practitioners from all countries confirm that conflicts of interest among industry representatives from one country are the rule rather than the exception. At the same time, coordination to ensure participants speak with one voice helps to establish support around a given standard proposal. In the West, such coordination is left to industry or to committees within private SDOs. While China's unity is often overestimated, in fields of national priority such as 5G, the Party-state indeed actively facilitates coordination. For the purpose of coordination, in 2013, the PRC founded the IMT 2020 (5G) Promotion Group, which comprises Chinese public agencies (Ministry for Industry and Information Technology, Ministry of Science and Technology and the National Development and Reform Commission), research institutes (Beijing University of Posts and Telecommunications) as well as all sorts of Chinese tech companies (Chen and Kang 2018).

When it comes to de facto standardisation outside of existing SDOs, China has exploited similar mechanisms that we are familiar with in the West. Large companies, package deals in which technical standards are adopted alongside favourable financial conditions and the creation of long-term liabilities are practices that the PRC has not invented. However, in contrast to most Western cases, the party-state has been actively engaged in the creation of large firms (Lardy 2019), state-owned banks have provided the funding for BRI infrastructure projects that spread Chinese technical standards and long-term liabilities in these projects often make recipient countries dependent from Chinese state-owned firms, not private companies. In short, while China's technical standardisation approach shows some similarities with Western practices and the externalisation of domestic structures resembles some Western tactics too, the PRC has adopted a much more state-directed approach. This contributes to two current trends: the politicisation and the fragmentation of technical standard-setting.

*Politicisation of technical standards.* Although historically a subject of power politics of states, the potential source of state power stemming from technical standards has largely been ignored by practitioners for the last few decades. The Chinese party-state's strategic approach to international technical standard-setting coupled with the emerging power competition over high technology is leading to a politicisation of the subject. Not least China's growing footprint could lead other, primarily developing countries to consider adopting a state-steered approach as it is practiced in the PRC (Rühlig and ten Brink 2021). Such politicisation alters the character of standardisation.

In *economic* terms, the politicisation incorporates a focus on the conditions upon which actors from different political entities get involved in international technical standard-setting. In strategic sectors such as key-enabling technologies, Chinese firms profit from party-state support. Hence, technical standardisation could be included in the West's drive to create a level playing field including sanction regimes.

Technical standards have further unfolded significant transformative force because once the world had agreed on a standard it was costly to change it. Complementary products and technologies would have needed to go through a process of adaptation generating switching costs. This relatively unquestioned character made international technical standards into an accepted part of international trade *law.* If technical standardisation will increasingly be seen through a political lens the "impartiality" attributed to standards could be undermined.

The politicisation of technical standards also directs the attention to cybersecurity implications of standard-setting. One example for the *political* dimension is that the US Department of Defence has issued concerns that China's strong presence in 5G standardisation could shift 5G technological development to focus on low-frequencies while US manufacturers have prioritised high-frequencies (mmWave). Some US experts argue that high frequencies provide a greater degree of security for wireless communications. Another concern is that US troops might need to rely on Chinese technology for their communications in overseas operations given Chinese strength in low frequency 5G technology (DIB 2019).

Finally, the technological character of standard-setting negotiations has long covered the *ideational* dimension of standardisation. Surely, technical standards have never been nonpolitical in substance. However, the recent politicisation could substantially change the process of standard-setting. The actors involved could pay more attention to ethnical, societal, and political underpinnings of different technological solutions. Hence, technical standardisation could turn into an arena of competition over values.

*Fragmentation of international standardisation.* As a result of the politicisation of technical standards standardisation could suffer from a divide into two camps. China could aim to develop a rivaling system of international standard-setting with the BRI serving as its steppingstone to outcompete established standardisation powers such as EU member states and the United States.

The potential divide into two distinct spheres of technical standards carries direct *economic* risks. One of the main benefits of technical standards is that they provide interoperability thereby facilitating international trade and harmonising technical necessities for market access. The potential to sell products on global markets is a driver for technological innovation. If companies have to design products in a distinct manner for different geographical areas, they suffer from a loss of efficiency. The increase in costs would hamper innovation.

Similarly, if there was no common understanding of what accounts as an international standard in the future, courts and other economic dispute settlement bodies could treat competing international standards as benchmarks for their considerations. If China establishes more dispute settlement mechanisms—as it has started to do under the BRI framework—a fragmentation of international *law* could be the result of competing international standards.

A bifurcated international standardisation would also facilitate *politically* relevant lock-in effects as described above. If standards are not global in scope, states depend on the supply of a potential adversary. In this context, several BRI railway projects financed by China's state-controlled financial institutions have raised concerns, such as the Jakarta-Bandung high-speed railway, the Abuja-Kaduna railway, the Ethiopia-Djibouti railway, and the China-Laos railway. All these projects are based on domestic Chinese technical railway standards, which means that the respective countries must rely exclusively on Chinese suppliers to maintain and further build out their railway networks.

Another widespread concern is that technical standardisation could turn into an arena of a new Cold War with two distinct blocks competing over values inscribed in technology. For example, China's expressed intention to invest in facial recognition standards (Xue Yujie 2019) and its proposal for a reformed standard internet protocol, referred to as New IP (Gross and Murgia 2020), have alarmed experts in Europe and the United States. The fear is that political and ethical preferences shaped by the political and societal framework of the party-state are incorporated in international standards.

## CONCLUSION

Technical standardisation is widely regarded as one of the main battlegrounds over technological leadership between the United States and China. This chapter explained why technical standardisation power is beneficial to the Chinese party-state in economic, legal, political, and ideational terms. Hence, China has shown significant efforts to increase its influence. These efforts have led to remarkable results though to varying degrees in different international SDOs. China has become an international technical standardisation power, but it is far from dominating standard-setting. Finally, China is—just like many other standardisation powers—externalising its domestic standardisation approach. The PRC has learned and adapted well-established practices but twisted them in order to fit the characteristics of China's political economy. This runs the risk of a further politicisation and fragmentation of technical standard-setting.

## NOTES

1. For a good overview including examples, see Väisänen, T. A. (2011). *Enforcement of FRAND Commitments under Article 102 TFEU: The Nature of FRAND Defence in Patent Litigation* (1st ed.). Baden-Baden: Nomos Verlagsgesellschaft mbH.

2. Footprint is defined as the ability to shape international technical standards.

3. Pohlmann, T. 2020. *Back To Basics Summer Webinar Part 2: SSOs, Patent Pools and Licensing* (Berlin: IPlytics, 2020), 11.

4. Author interviews with European engineers involved in the development of 5G. February–November 2019, several cities.

5. Author interviews with international standardisation representatives. October 2018–April 2020, several cities. For details, see appendix.

6. Information privately by the author from the German SDO, Deutsches Institut für Normung (DIN).

7. Calculations of the author based on data obtained privately from a government agency in an EU member states being involved in mobile network standardisation in 3GPP.

8. DiploFoundation. 2021. The geopolitics of digital standards: China's role in standard-setting organisations, issued December. Available from https://www.diplomacy.edu/resource/report-the-geopolitics-of-digital-standards-chinas-role-in-standard-setting-organisations/ [18 November 2022].

9. Calculations of the author based on data obtained privately from a government agency in an EU member states being involved in mobile network standardisation in 3GPP.

10. Author interviews with non-Chinese standardisation representatives. October 2018–April 2020, several cities. For details, see appendix.

11. Author interview with a senior representative of a European standardisation organisation, November 2018, Brussels.

12. Author interview with a senior representative of a European standardisation organisation, April 2019, Stockholm.

13. Author interview with junior representative of a European standardisation organisation, December 2018, Brussels.

14. Author interviews with Chinese standardisation representatives. March 2019–March 2020, several cities. For details, see appendix.

15. Author interview with a senior national ministry official, October 2019, Qingdao.

16. "标准联通共建'一带一路'行动计划(2018–2020年)," SAC, 2018, accessed 2018-10-26, http://www.sac.gov.cn/zt/ydyl/bzhyw/201801/t20180119_341413.htm.

17. Based on information obtained by the author from the Standards Administration of China (SAMR/SAC).

18. Based on privately obtained documents as well as author interviews with Chinese standardization officials. September–October 2019, several cities.

## REFERENCES

Arthur, W. B. 1989. Competing technologies, increasing returns, and lock-in by historical events. The economic journal 99(394), 116–31.

Atlantic Council. 2021. The China Plan: A Transatlantic Blueprint for Strategic Competition, issued March. Available from: https://www.atlanticcouncil.org/wp-content/uploads/2021/03/The-China-Plan-A-Transatlantic-Blueprint.pdf [10 April 2021].

Bonardi, J. P., and Durand, R. 2003. Managing network effects in high-tech markets. *Academy of Management Perspectives* 17(4), 40–52.

Brookings Institution. 2021. Huawei Meets History. Great Powers and Telecommunications Risk, 1840–2021, issued March. Available from: https://www.brookings.edu/wp-content/uploads/2021/03/Huawei-meets-history-v4.pdf.

Busch, L. 2011. *Standards. Recipes for Reality* (Cambridge: MIT Press).

Chen, S-Z. and Kang, S-L. 2018. A Tutorial on 5G and the Progress in China. *Frontiers of Information Technology & Electronic Engineering* 19(3), 309–21.

Defence Innovation Board (DIB). 2019. The 5G ecosystem: Risks and opportunities for DoD, issued 4 March. Available from: https://apps.dtic.mil/sti/citations/AD1074509 [18 November 2022].

Deron, Laure G. 2020. Chinese Standards and the New Industrial Markets. *Institut Rechereche Stratégique de l'École Militaire Research Paper* 98, 2–15.

Gargeyas, A. 2021. China's 'Standards 2035' Project Could Result in a Technological Cold War. The Diplomat, 18 September. Available from: https://thediplomat.com/2021/09/chinas-standards-2035-project-could-result-in-a-technological-cold-war/ [11 September 2022].

DiploFoundation. 2021. The geopolitics of digital standards: China's role in standard-setting organisations, issued December. Available from https://www.diplomacy.edu/resource/report-the-geopolitics-of-digital-standards-chinas-role-in-standard-setting-organisations/ [18 November 2022].

Eisenstark, R. 2018. Why China and the US are fighting over 5G technode, 30 March. Available from: https://technode.com/2018/03/30/5g/ [4 November 2019].

Farrell, J., and Saloner, G. 1985. Standardization, Compatibility, and Innovation. *The RAND Journal of Economics* 16(1), 70–83.

Graz, J-C. 2019. Standards as Regulation. In *The Power of Standards: Hybrid Authority and the Globalisation of Services*, Cambridge: Cambridge University Press, 86–113.

Gross, A. and Murgia, M. 2020. China and Huawei Propose Reinvention of the Internet. Financial Times, 27 March. Available from: https://www.ft.com/content/c78be2cf-a1a1-40b1-8ab7-904d7095e0f2 [30 April 2020].

Hallström, K. T. 2004. *Organizing International Standarization. ISO and the IASC in Quest of Authority*. Cheltenham: Edward Elgar.

International Organization for Standardization (ISO). 2018. My ISO job: What delegates and experts need to know, issued n.d. Available from https://www.iso.org/files/live/sites/isoorg/files/archive/pdf/en/my_iso_job.pdf [23 April 2021].

Krislov, S. 1997. *How nations choose product standards and standards change nations*. Pittsburgh: University of Pittsburgh Press.

Lardy, N. R. 2019. *The State Strikes Back. The End of Economic Reform in China?* New York: Columbia University Press.

Lee, H. and Oh, S. 2006. A standards war waged by a developing country: Understanding international standard setting from the actor-network perspective. *The Journal of Strategic Information Systems* 15(3), 177–95.

Medin, M. and Louie, G. 2019. The 5G Ecosystem: Risks & Opportunities for DoD. Defence Innovation Board, 1–31.

OECD Working Party of the Trade Committee. 1999. Regulatory Reform and International Standardisation TD/TC/WP36/FINAL, issued 29 January. Available from: no link.

PCR. 2015. Action Plan to Connect "One Belt, One Road" Through Standardization (2015–2017), issued n.d. Available from https://www.followingthemoney.org/wp-content/uploads/2017/06/2015_Leading-Group-for-the-BRI_Action-Plan-to-Connect-BRI-through-Standardization-2015-2017_E-1.pdf.

Pohlmann, T., Blind, K., and Heß, P. 2020. Studie zur Untersuchung und Analyse der Patentsituation bei der Standardisierung von 5G. *Study on behalf of the German Bundesministerium für Wirtschaft und Energie*, Berlin.

Pop, V., Hua, S. and Michaels, D. 2021. From Lightbulbs to 5G, China Battles West for Control of Vital Technology Standards. *Wall Street Journal*, 8 February. Available from: https://www.wsj.com/articles/from-lightbulbs-to-5g-china-battles-west-for-control-of-vital-technology-standards-11612722698 [17 February 2021].

Rühlig, T. 2020. Technical standardisation, China and the future international order. A European perspective. *Heinrich Böll Stiftung*, 6–35.

Rühlig, T. and Björk, M. 2020. What to make of the Huawei debate? 5G network security and technology dependency in Europe. *UI Paper 1*, 4–37.

Rühlig, T. N. and ten Brink, T. 2021. The Externalization of China's Technical Standardization Approach. *Development and Change* 52(5), 1196–1221.

SAC. 2017. 'Zhōnghuá rénmín gònghéguó biāozhǔnhuàfǎ' (Standardization Law of the People's Republic of China), issued 08 November. Available from: www.sac.gov.cn/sbgs/flfg/fl/bzhf/201711/t20171108_318652.htm [5 February 2021].

SAC. 2018. "'标准联通共建'一带一路'行动计划(2018–2020年" (Action Plan for Standard Unicom to Build the "Belt and Road" (2018–2020)), issued 19 January. Avaiable from: http://www.sac.gov.cn/zt/ydyl/bzhyw/201801/t20180119_341413.htm [26 October 2018].

Schilling, M. A. 2002. Technology success and failure in winner-take-all markets: The impact of learning orientation, timing, and network externalities. *Academy of management journal* 45(2), 387–98.

Schneiderman, R. 2015. International Standards Development Organizations Defined. In *Modern Standardization. Case Studies at the Crossroads of Technology, Economics and Politics*. Piscataway: IEEE Press, 2015, 253.

Seaman, J. 2020. China and the new geopolitics of technical standardization. *Notes de l'Ifri* 34, 3–31.

SFRC Democratic Staff. 2020. The New Big Brother—China and Digital Authoritarianism, issued 21 July. Available from: https://www.govinfo.gov/content/pkg/CPRT-116SPRT42356/pdf/CPRT-116SPRT42356.pdf.

Shalal, A. 2020. Biden Adviser Says Unrealistic to 'Fully Decouple' from China. Reuters, 22 September. Available from: https://www.reuters.com/article/us-usa-trade-china-biden-idUSKCN26D1SM [12 April 2021].

Shi-Kupfer, K. and M. Ohlberg. 2019. China's Digital Rise. Challenges for Europe. *Merics* 7, 7–55.

Strumpf, D. 2019. Where China Dominates in 5G Technology. WSJ. 26 February. Available from: https://outline.com/dVsKLJ [13 April 2019].

Suttmeier, R. P., Yao, X. and Tan, A. Z. 2009. Standards of power? Technology, institutions, and politics in the development of China's national standards strategy. *Geopolitics, History, and International Relations* 1(1), 46–84.

The 'China, Inc.' Challenge to Cyberspace Norms. Hoover Institution, n.d. Available from: https://www.hoover.org/sites/default/files/research/docs/williams_webreadypdf1.pdf.

Timmermans, S. and Epstein, S. 2010. A world of standards but not a standard world: Toward a sociology of standards and standardization. *Annual review of Sociology* 36(1), 69–89.

Väisänen, T. A. (2011). *Enforcement of FRAND Commitments under Article 102 TFEU: The Nature of FRAND Defence in Patent Litigation* (1st ed.). Baden-Baden: Nomos Verlagsgesellschaft mbH.

Williams, R. 2018. The 'China, Inc.'Challenge to Cyberspace Norms. Hoover Institution, n.d. Available from: https://www.hoover.org/sites/default/files/research/docs/williams_webreadypdf1.pdf.

World Trade Organization. 2012. Trade and public policies: A closer look at non-tariff measures in the 21st century, issued n.d. Available from https://www.wto.org/english/res_e/booksp_e/anrep_e/world_trade_report12_e.pdf.

Wu, M. 2016. The China, Inc. challenge to global trade governance. *Harvard International Law Journal* 57(2), 261–324.

Xinhua. 2021. 两会受权发布□中华人民共和国国民经济和社会发展第十四个五年规划和2035年远景目标纲要 (The 14th Five-Year Plan for National Economic and Social Development of the People's Republic of China and Outline of Long-term Goals for 2035), issued 13 March. Available from: http://www.xinhuanet.com/2021-03/13/c_1127205564.htm [3 April 2021].

Xue Yujie. 2019. 27 Companies Drafting China's First National Facial Recognition Standard. Sixth Tone, November 27. Available from: http://www.sixthtone.com/news/1004893/27-companies-drafting-chinas-first-national-facial-recognition-standard [04 April 2020].

Yan, Xuetong. 2020. Bipolar Rivalry in the Early Digital Age. *Chinese Journal of International Politics* 13(3): 313–41.

*Chapter 6*

# China and Global Data Transfers

## *Implications for Future Rulemaking*

Hunter Dorwart

Global data transfers form a cornerstone of the digital economy. Without them, digital services and trade would cease to function, and individuals would lose the ability to communicate across borders through the Internet (Casalini and Gonzalez 2019). Measuring the value of global data flows remains challenging, but recent research demonstrates that they could contribute up to $11 trillion to global GDP by 2025, reduce export costs and transaction times by over 60 and 30 percent respectively, and accelerate new opportunities for businesses across the globe (UNCTAD 2021; Casalini and Gonzalez 2019; Alphabeta 2019).

Due to their importance, data flows have become a contentious issue between governments and frustrated efforts to create international consensus around the rules of digital trade and cyberspace (Kuner 2011). The Court of Justice of the European Union (CJEU) has held twice (in its 2015 and 2020 *Schrems* rulings) that data transfers from the EU to the United States violate the European Union General Data Protection Regulation (GDPR) because US law does not afford an essentially equivalent level of protection to individuals in the European Union (EU) as does the GDPR. In the United States, concerns over the Chinese Communist Party's (CCP) access to data has led the US government to ban some Chinese companies (e.g., Huawei, Hikvision) from the market and convince other states to do the same.

Because of this, disagreements over data transfers underscore larger debates about the future of the Internet and the prerogative of nation states to assert different values over the digital realm (Mueller 2017; Hummel 2021). Restricting data transfers is one mechanism states use to exert jurisdiction over cyberspace, which makes such restrictions intertwined with data sovereignty, a concept increasingly promoted by states to justify their authority over technology and data (Chander and Sun 2021; Gao 2022).

China's rise as a cyber power and its growing global influence over digital technologies has fueled these debates. By 2025, estimates indicate that China's internet population will reach 1.14 billion, the largest in the world, while e-commerce now accounts for 35 percent of total retail sales in the country (and 40 percent globally) with a market size expected to reach $5.6 trillion in the upcoming years (Zhang and Chen 2019; Jiang 2020). Some of the largest Internet companies from China, including Alibaba, Tencent, and Jingdong (JD), and the country now hosts nine of the top twenty global internet firms by market value (Liu 2021). These firms have increasingly gone global: TikTok has become the world's most downloaded mobile app while Huawei continues to dominate the telecommunications equipment market (McAuliffe 2022; Pongratz 2020).

At the same time, the People's Republic of China (PRC) has for years promoted its own approach to data governance—what it calls 'cyber sovereignty' (*wangluo zhuquan*)—that resonates with the concept of data sovereignty (Creemers 2020). Relatedly, many scholars and policymakers have attributed the emergence of an array of governance tools and values to China including online censorship, state surveillance, and even digital authoritarianism. While each of these trends predates the rise of China as a cyber and technology power, widespread perceptions continue to associate the country with legitimising and enabling state control over cyberspace and information (Segal 2020).

With respect to data flows, a core part of Chinese practice involves data localisation, or the practice of forcing organisations to use, store, and process data within a country's sovereign territory (Chander and Le 2015). China has in recent years solidified a far-reaching data localisation framework that requires entities to process certain data locally and receive government approval for overseas transfers. This framework raises numerous questions about the future rules of global data transfers. What are China's goals beyond its own domestic priorities and how may they affect international discourse and practice? Does China want to rewrite the rules of data transfers to better favour Chinese interests abroad and if so how and to what extent will it do this? Has China already influenced international rulemaking or displaced other mechanisms?

This chapter attempts to address these questions by (1) analysing China's strategic goals with respect to global data transfers and (2) exploring the extent to which the PRC has realised its objectives. It argues that Chinese policymakers want to enshrine two goals—what this chapter calls the *twin aims*—into the foundation of global data transfers. These aims involve facilitating data flows across borders but in a way that ensures security and regime stability.

While China's overarching objective is clear, its strategy to translate the twin aims into international rules and principles remains inchoate. Not only are there information gaps in China's preferred outcome, but current evidence also suggests that the roadmap for practical implementation remains fragmented. The country's leaders have stressed the need for China to help formulate international rules with partners, and have even launched global initiatives to start this, but have produced few results that would meaningfully structure international data transfers *outside of China* or influence other countries to adopt China's preferred approach.

Nonetheless, through its *data ordering* (a combination of regulatory rule-making and technology practice), the PRC has produced three *spill-over effects* that have formed a multi-dimensional foundation through which China may influence global data transfer rules in the future. Data ordering is related to data governance, which refers to the regulatory and technical tools that governments employ to govern information and technology but is different as it emphasises the practice of Chinese administrative institutions to instil order through means that dodge conventional legal concepts and vocabulary (Clarke 2020).

Each spill-over effect contains key shortcomings that complicate measuring China's current impact on data transfers rulemaking. Part of the difficulty lies in determining whether China wants to construct international rules and exert regulatory leadership or whether officials wish to assert influence only to favour a limited, domestic interest. In other words, translating China's domestic aims into international goals with respect to data transfers involves taking a conceptual leap in an environment where existing evidence does not always point to an easy solution. As of writing, the answer remains relatively unclear. Yet the shape of China's data ordering has come into form through the following spill-over effects:

1. The solidification of China's *data governance regime*, and in particular its approach to data localisation, has shaped corporate expectations and compliance practice around cross-border data flows. Yet despite the far-reaching implications of this system, its influence may be limited to China's domestic market.
2. Beyond this, China has begun to legitimise its *norms and principles* regarding control over data with other governments and plurilateral institutions. It has done so through a combination of soft law and participatory diplomacy, but the impact of this on data transfers remains hard to measure in the absence of any concrete agreement or rules.

3. Chinese *technology exports*, and in particular data centres, shape corporate governance and exert a bottom-up force on data transfer rules. Government plans to create data 'customs' hubs to facilitate data flows securely across borders may serve as a model that could scale in foreign jurisdictions. However, this impact may be limited to data flows in and out of China, rather than between countries.

This chapter traces the development of China's strategy and the three spill-over effects by analysing government policy documents, administrative regulations, and corporate public data in China. It also relies on interviews and conversations with Chinese lawyers, compliance teams, and data security researchers to explain how companies have responded to China's data transfers framework and its impact on global data flows. While it focuses primarily on corporate transfers, many of this chapter's arguments apply to the important issue of government access to data, although relatively underexplored.

In highlighting these vectors, this chapter attempts to lay the foundation for future research and draw attention to methodological challenges that face grasping the extent of China's rise as a cyber power. Any analysis of China's technology strategy and capabilities must confront gaps in information and recognise that in the absence of additional data, researchers must not let their assumptions finish the explanation. This chapter attempts to highlight these gaps while drawing attention to known trends.

## CONCEPTUALISING CHINA'S STRATEGIC GOALS FOR GLOBAL DATA TRANSFERS

Over the past few years, China's strategic goals for global data transfers have come into view, but their concrete path for implementation remains unclear. At the heart of this uncertainty lies the *twin aims* of balancing the free flow of data across borders for commercial interests with a restrictive approach that prioritises security. Put simply, China wants to realise both goals—it wants a world where data governance principles facilitate digital development at home and abroad, but also one where it can develop comprehensive policy tools like data localisation to uphold security. How and through what tools China will shape this world remains unanswered.

This section attempts to provide an account of the *intention* behind China's strategic goals with respect to global data transfers by first situating the twin aims within the broader context of China's cyber governance values. It argues that the *orderly flow of data* has emerged as a unique concept that reflects the need to prioritise national security over other interests. Next, this section highlights emerging ambiguities in China's strategy to translate its domestic goals into international rules. While China has promoted cooperation to create international rules, its efforts so far have left many questions unanswered.

## The Orderly Flow of Data and the Twin Aims of Global Transfers

Chinese authorities have long viewed data as necessary to develop the digital economy and make China more self-sufficient in technological innovation and growth (Arsene 2018; Gu & Lundvall 2006). Chinese leaders launched a national data strategy in 2014, followed by a series of government work reports dedicated to exploring the potential use cases of big data (CAICT 2019). In 2015, the State Council issued an action plan to promote data development, one of the earliest top-down strategic planning documents for data flows, which called for the construction of integrated data systems and the informatisation of a whole range of strategic industries and government functions (State Council 2015). Implementing the big data strategy became a key goal in the 13th Five-Year Plan (FYP 2016–2020) and received support directly from Xi Jinping in 2018 when he stressed the need to 'promote the deep integration of the Internet, big data, and artificial intelligence with the real economy' (State Council 2016; Qiushi 2021). In April 2020, a guideline on improving market-based allocation of production factors listed data as the fifth basic factor of production along with land, labour, capital, and technology (Central Committee and State Council 2020). At the core of this perspective is treating data like a resource that can be leveraged for economic growth through market mechanisms (Liu 2021).

However, Chinese leaders also early on recognised the unprecedented risks the digital economy brings for China's national security, political rule, economic independence, and social stability (Lu 2014). Part of this concern emanated from China's turbulent historical relationship with foreign powers and its sensitivity to social stability and political control—an outlook shaped by its experience as a postcolonial state forged in the context of war (Mühlhahn 2019). Another part originated from it being a relative latecomer to the information communication technology (ICT) revolution and its lack of self-sufficiency due to its early reliance on American technologies (Yang 2014).

To bolster security online and respond to criticisms of its early censorship and state control over the Internet, Chinese scholars and officials formulated the concept of cyber-sovereignty, a set of overarching governance principles which seeks to justify state control of the Internet in the name of social order and stability (Arsene 2016). At least since 2010, the Chinese government has referred to the concept to describe China's normative position on cyber governance. It has done so both in internal strategy documents and through its participation in multilateral institutions (SCIO 2012; MFA 2012; MFA 2015; MFA 2017). When it comes to the governance of data, policymakers have applied a similar rationale and created administrative and technical tools to censor the flow of information online and across borders (Zheng 2020).

Taken together, the need to ensure data flow for economic development while restricting access to preserve security forms the twin aims of the Chinese government to data flows. The twin aims raise important implications for cross-border data transfers and China's strategic goals for international data governance. On the one hand, policymakers connected to commercial institutions recognise that the free flow of data across borders is paramount for Chinese businesses, especially as they expand into foreign markets. To this end, they have initiated a series of pilot programs designed to maximise digital trade through 'data free trade zones' and explored cross-border governance on the local level through data exchanges (MOFCOM 2020; Lu 2020).

On the other hand, security-focused policymakers view the unregulated flow of data across borders as a threat to China's sovereignty and public order. In recent years, these policymakers have developed comprehensive data localisation rules designed to minimise security risks and maximise state control over data. Authorities have increased their enforcement of these rules, with Didi Chuxing, Zoom, and the China National Knowledge Infrastructure (CNKI) being notable examples of companies that have had their operations halted or severely restricted due to alleged violations (CAC 2022; CAC 2022a).

To address both aims, Chinese policymakers describe the ideal for cross-border transfers as the *orderly flow of data* (*shuju youxu liudong*). Increasingly, numerous policy documents have used the term, including the Central Cybersecurity and Informatisation Commission (CCIC) 14th Five-Year Plan for Informatisation and pilot programs under the Ministry of Commerce (CCIC 2021; MOFCOM 2020). The Cybersecurity Law (CSL), the Data Security Law (DSL), and numerous regulations issued by the Cyberspace Administration of China (CAC) also refer to the orderly flow of data in the context of transferring data overseas (NPC 2021; NPC 2021a; CAC 2022b).

But what does it mean for data to flow in an orderly manner? The CAC predominantly uses the term in relation to what it describes as the 'lawful' and 'free' flow of data, but always in the context of regulations that add compliance requirements for organisations to transfer data abroad. Because of this, the orderly flow of data at a minimum implies that the transmission of data will not harm China's national security, public order, or political interest. It likewise implies that data shall not flow unless certain conditions are met, including compliance with Chinese law.

Chinese policymakers are still working out the balance between commercial and security interests when it comes to data flows, which makes arriving at a systematic definition of the concept somewhat of a moving target. The twin aims have often produced tension within China's political and administrative institutions, and in ways that have not always been easy to resolve (Boullenois 2021). Such tension has involved battles over regulatory turf, power-sharing between new and old bodies such as the CAC and the Ministry of Public Security (MPS), and concerns about the position of tech companies within China's larger digital economy (Creemers 2021; Zhang 2016). The Chinese government has yet to resolve many details of the twin aims, which has created even more ambiguity in the context of international rulemaking.

## Ambiguities in China's Strategy for Shaping International Transfer Rules

How Chinese leaders will attempt to influence international data transfer rules in a way that reflects the twin aims remains unclear. At least since 2015, policymakers and strategists have promoted the need for China to build a 'community of common destiny for cyberspace' by calling on other countries to enshrine four principles into the international Internet governance system: respect for sovereignty, peace and security, openness and cooperation, and order (Lu 2016). Leaders reiterated these principles in the 2017 International Strategy of Cooperation on Cyberspace, which emphasised the priority of developing a rules-based global governance system to 'promote orderly information flows' (MFA 2017).

Through these principles, China has made legitimising its security-oriented approach to the Internet a key strategy on the international stage (Wu 2021). However, except for a few limited instances, the country's leaders have provided less guidance on what rules should govern data flows beyond respecting sovereignty and how or through what fora China should go about creating those rules. In other words, there are conceptual problems when translating China's domestic goals on data flows into international aims (Zheng 2022; Bergsten 2022).

Despite this, there are signals where China may be heading. Some Chinese scholars within ministry think tanks argue that China is following a tiered approach that starts first with solidifying China's data regulations and improving cross-border data flow pilots before engaging in international negotiations or digital trade deals to craft lasting rules (Zheng 2022). As discussed below, under this framework China has completed the first stage and is in the process of operationalising the second while initiating the third and fourth (Weng and Song 2022).

Chinese ministries, administrative bodies, and other lawmakers have also indicated their preference for rulemaking, albeit in a general manner (Hong 2020; Zheng 2020; Huang 2020). For instance, both the DSL and the PIPL speak of the need for China to 'actively participate in the formulation of international rules' with respect to international data transfers, but do not specify what rules should be created (Wu 2021).

More concretely, the Chinese government has made data flows a key priority in two types of international settings: 1) negotiations in trade agreements that contain digital services chapters; and 2) plurilateral institutions and initiatives that involve data regulation. While China has increased its participation in both settings, efforts to create binding rules have so far produced mixed results. This has generated scepticism as to China's true strategic aim with both settings. Such ambiguity raises questions about what China's leaders wish to accomplish with data transfers.

## Data Flows and Digital Trade Agreements—Rationales and Limitations

China's track record on creating data transfers rules through free trade agreements (FTA) remains limited (Gao and Schaffer 2020). The FTAs it has signed or applied to join lack comprehensive rules in a manner that would solve the most pressing global issues on data flows (Mishra 2021; Voss 2020; Yakoleva & Iron 2020). Of the country's recent bilateral FTAs, only its agreements with Korea and Australia contain standalone chapters on e-commerce and both carry limited obligations for data protection (Gao 2018).

On a regional level, China conditioned its entry into the Regional and Comprehensive Economic Partnership (RCEP) agreement on the grounds that it could exempt its data localisation rules from the agreement's general prohibition against transfers restrictions. However, this exemption doesn't create any positive rules that would bring legal certainty for other countries (Gao 2018). A similar pattern exists in China's participation in the World Trade Organisation (WTO). Through its commitment to the E-Commerce Joint Statement Initiative (JSI), which is an effort to create new plurilateral rules on e-commerce including data transfers, China has committed to advancing negotiations, but has routinely asserted that security should serve as a precondition for data flows across borders, a position it has adopted in other multilateral fora (Erie and Streinz 2021).

One notable exception concerns China's interest in joining the Comprehensive and Progressive Agreement for Trans-Pacific Partnership (CPTPP), the resuscitated version of the Trans-Pacific Partnership (TPP) that the Obama administration spearheaded to set the rules for digital trade in Asia. While the United States later withdrew from the initiative, its other signatories have revived it, making the agreement one of the largest ongoing trade deals in negotiation (Williams and Sutherland 2021). The CPTPP contains language to prohibit countries from adopting a data localisation model like the PRC's. It goes further than RCEP by requiring members to demonstrate that restrictions on data flows meet a legitimate public policy objective, are not discriminatory or arbitrary, and are proportionate and necessary (Kong and Tong 2021). Chinese experts assert that the country's data governance laws are compatible with CPTPP's conditions, despite widespread scepticism (Hong 2020). However, as of writing, the status of China's ascension to the agreement is still ongoing.

## The Global Data Security Initiative (GDSI)

China has also advanced rulemaking around data transfers in other plurilateral institutions. It has increased its participation in well-known internet governance bodies like the International Telecommunication Union (ITU) and even launched its own international institutions such as the World Internet Conference (WIC) to promote cooperation on cyberspace governance (see Section III[B][2]). As discussed below in more detail, grasping China's intentions with respect to data transfers through its engagement in these bodies remains difficult, as data flows are only one issue among many others.

The initiative that Chinese leaders have backed that explicitly touches upon data transfer rules is the Global Data Security Initiative (GDSI, *quanqiu shuju anquan changyi*). Announced in September 2020, the GDSI attempts to provide a framework for countries to cooperate on issues related to cross-border data flows, including law enforcement access to data and information security (Park 2022). It proposes three principles (multilateralism, safety, and fairness) and eight action items to structure how governments should approach solving issues related to data flows (MFA 2020). These action items involve broad positive commitments such as protecting personal information, complying with domestic laws, and addressing law enforcement access to data through judicial assistance, as well as negative prohibitions including promises not to directly access data located in other countries, set up backdoors in products or services, or even force countries to localise data.

The high-level principles and relative openness of the GDSI suggest that Chinese policymakers view cross-border transfer rules as an avenue to build a larger framework for the global digital economy. Since 2020, the GDSI has been promoted at various levels in the Chinese government. Xi Jinping mentioned the initiative in three separate multilateral summits in 2020: the Shanghai Cooperation Organisation (SCO) Summit, the BRICS Summit, and the G20 (Park 2022). Chinese ministries have made the GDSI a key part of its bilateral and regional engagement on cyber issues including in fora like the China-Arab Data Security Cooperation and the China-ASEAN Digital Governance Cooperation.

The GDSI has received support, albeit in a general and limited manner, from numerous countries around the world especially in the Global South. However, little information beyond terse readouts and press releases of this engagement exists. Indeed, the GDSI's role in larger plurilateral fora remains unclear—the initiative proposes principles that other countries may agree to but leaves the work of filling in the details to other cooperative efforts (Park 2022). China may prefer to do this on a bilateral basis, as its recent cybersecurity agreements with Thailand and Indonesia suggest, or it may increase its efforts in other institutional settings (CAC 2022c; CAC 2022d).

## THE SPILLOVER EFFECTS OF CHINESE DATA ORDENING—PRESENT AND FUTURE INFLUENCE

While the pathway for China to accomplish its twin aims remains fragmented, the PRC has nonetheless already influenced the global debate around data flows, internet governance, and cross-border transfers. This influence is not limited to formal rulemaking but extends to other practices such as governance norms and design choices in contracts. These different vectors—hard law, soft law, corporate governance—overlap with each other but exist on different conceptual planes.

As a result, measuring the effects of each requires different methodological tools. For instance, focusing exclusively on whether Chinese data protection law will become a global de facto standard like the GDPR risks ignoring how the country's technology exports may shape corporate governance practices in recipient jurisdictions (Erie and Streinz 2021). Likewise, an emphasis on soft law and other non-binding mechanisms like standards may ignore the concrete impact and cost China's localisation rules have already had on global technology firms.

The lack of a consistent framework that addresses each of these dimensions has challenged evaluating China's influence on global data transfers. This section attempts to address this gap by identifying three spill-over effects of Chinese *data ordering* that have and will continue to impact debates around data flows. Data ordering involves a combination of law and technology to create certainty and norms over the management of data. It includes the participation of regulatory authorities that implement data regulations and technology companies that design and export commercial products. Notably, data ordering reflects that some of China's administrative practices differ considerably from other legal systems (Clarke 2020).

First, the solidification of China's data governance regime has shaped international expectations with respect to data localisation, influencing both how private and public sector actors understand global transfers. This solidification involves not only the adoption of regulations that mandate localisation, but also the promulgation of alternative transfer mechanisms such as standard contractual clauses (SCCs) and cross-border certification systems. Notably, while China's influence on global data transfers has increased, the impact of its law on formal rulemaking or international compliance standards likely faces limitations, as China's data protection framework is mostly relevant for companies engaging in the Chinese market.

Second, Chinese data ordering may influence other countries by shaping norms and expectations around data sovereignty, particular in their data governance choices. One area where this is most visible is China's participation in international governance institutions and the direct dialogue with foreign governments and their officials in such fora. However, measuring the effect of this influence on a case-by-case basis remains challenging, while China's participation in international institutions has so far produced mixed results.

Third, Chinese companies have shaped global data flows discourse by exporting governance through technology. In particular, technical tools and programs designed to manage big data are beginning to affect recipients' approaches to data flows and their contractual relationships with Chinese vendors. One notable example concerns the proliferation of data centres and their ability to facilitate data transfers through compliance services—a phenomenon already witnessed in Hong Kong's planned data 'customs' hub. Policymakers may attempt to scale this and other data exchanges into other jurisdictions, but the impact faces a similar limitation to the first effect in that it may only influence transfers to and from China, and not between third-party countries outside of China.

## China's Data Transfers Framework—Existing Capabilities and Influence

### *The Solidification of China's Data Localisation Architecture*

As stated above, Chinese policymakers have stressed the need for China to develop a coherent framework for data transfers before actively participating in international rulemaking (Wu 2021;Zheng 2020). Over the past few years, lawmakers have made progress in formulating an institutional design for cross-border data flows that contains baseline localisation requirements in some scenarios. While still incomplete, this framework requires certain

**Table 6.1. Three Spillover Effects of Data Ordering**

| *Description of Effect* | *Vector* | *Application* | *Limitations* |
|---|---|---|---|
| Solidification of data localization framework | Hard law | Influences corporate compliance expectations, operationalizes data localization, exerts sovereignty. | Primarily impacts entities engaging with the Chinese market. Impact on the rest of the world is limited. |
| Transnational principles and participatory diplomacy | Soft law | Legitimises norms and principles for other countries to follow. Sets agenda for plurilateral institutions and other government fora. | Difficult to measure legislative influence on other countries. Rulemaking in plurilateral institutions is limited, more principle-based than outcome-based. |
| Technology exports and data hubs | Corporate governance | Influences organizational choices of foreign firms. Data hubs may scale to facilitate security compliance and lower corporate costs. | Influence may be limited to data flows to and from China and not between entities outside of China. Connection between commercial technology exports and data centres is hard to measure. |

businesses to process their data locally within the territory of the PRC and receive mandatory security assessment (i.e., government approval) for overseas transfers.

In other circumstances, organisations must choose an enumerated transfer mechanism before sending data overseas, such as using a SCC or receiving certification for the transfer. The current system reflects a need to ensure the orderly flow of data, but also indicates the strong position of national security within that need, given that the threshold triggering localisation seems very low.

The development of this framework has not been straightforward and easy and has often reflected the changing attitudes of lawmakers to localisation and data security (Hong 2017). With the promulgation of the Cybersecurity Law, policymakers took a step towards localisation, but struggled to implement the rules on the ministerial level (CAC 2017; CAC 2019). The compilation of the Chinese Civil Code in 2020 and the adoption of the PIPL and the DSL in 2021 marked a new stage in the evolution of China's data transfers framework, which reflected compromises between competing interests and ministries (Creemers 2021).

Both laws concretise China's approach to data transfers, which revolves around two conceptual pillars: (1) the type of entity processing the data and (2) the type of data being processed. Article 40 of the PIPL reaffirms the need for critical information infrastructure operators (CIIOs) to process data locally—an obligation under the CSL—but also extends this requirement to data controllers that process a certain 'volume' of personal information. In 2022, the CAC adopted a regulation that defined this volume threshold at 1 million individuals, set forth a restriction on transferring 'important data' out of China for all organisations, and clarified the process for obtaining a security assessment (CAC 2022b). The regulation also stipulates that controllers who cumulatively transfer personal information of 100,000 individuals or sensitive information of 10,000 individuals must also localise their data and receive a security assessment.

Through this approach, officials within the CAC have indicated a strong preference for restricting certain data flows on the basis of protecting national interests such as security, public order, and safety (Huang 2020). But they have also tried to design the framework to leave open the possibility of enabling data flows for business purposes (MOFCOM 2020). The PIPL recognises the possibility of China entering a bilateral or plurilateral agreement on data transfers, which indicates a willingness of Chinese leaders to negotiate and shape international rules. As of writing, China has not formulated such an agreement, but has committed to working with other countries in principle (Hong 2020). During the Data Governance Forum in 2022, a mutual recognition treaty between China and Singapore was proposed, but details so far remain scant (MLex 2022).

## *Measuring the Impact of China's Data Transfers Framework Abroad*

The solidification of China's data transfers regime has already had a significant impact on how businesses, policymakers, and academics view global data transfers. This impact can be measured in three aspects. First, China's data protection architecture represents a unique model that cannot be reduced to falling somewhere between the United States and the EU approach to data transfers (Zheng 2020). The emergence of this model has spurred scholarship both within and outside of China to explain the law's key features on its own terms, which in turn has shaped how many in the profession view global data protection trends.

Second, legal teams in some of the largest foreign tech firms are beginning to modify their global compliance programs in light of China's model.

Compliance officers now consider China a unique regulatory market that demands a fresh conceptual approach (Huang 2020). As a result, many in the field associate China with enabling localisation and refer to the country when discussing global data protection trends or highlighting vicissitudes in cross-border transfers.

Third, China's model contains its own version of transfer instruments commonly used in other jurisdictions. For example, in 2022 regulators released a draft cross-border certification standard and draft SCCs, both of which are specified under Article 38 of the PIPL (TC260 2022; CAC 2022e). These mechanisms theoretically offer organisations an alternative path for transfers that does not rely on a security assessment (i.e., government-led) but instead utilises contractual law and third-party auditing (i.e., market-oriented). The promulgation of Chinese SCCs in particular is notable because of their widespread use by companies in other jurisdictions, particularly in the United States and the EU, and their connection with the ongoing transatlantic transfers debate (Zanfir-Fortuna 2021). Additionally, both instruments may help scale China's regulatory influence beyond its domestic market, especially if Chinese tech companies begin to use them for their subsidiaries located in countries outside of China.

### *Limitations of Translating Domestic Law into International Rules*

While Chinese data protection law is beginning to shape international opinions on data transfers, it remains unclear to what extent it will continue to do so or whether that influence will amount to meaningful global rulemaking. To be sure, China's current model has already impacted private companies and in part bolstered global trends towards data sovereignty (Erie and Streinz, 2021). Although not the only or the first country to do so, China's balance of asserting jurisdictional control over data while promoting commercial interest has resonated both positively and negatively with other policymakers, academics, and businesses (Nanni 2020).

However, translating this influence into global rulemaking remains limited and challenging. On the one hand, China's regulatory capacity for data governance is nascent and although its transfer model aspires to sophistication, many aspects of the framework are still ambiguous and undefined (Creemers 2021). In recent years China has spent considerable resources to reterritorialise data and apply its domestic law beyond the country's borders, yet the application of its framework abroad has been variegated and fragmented. In other words, Chinese law is nowhere near setting a de facto global standard like the GDPR (Bradford 2020).

Indeed, many companies do not favour adopting the Chinese approach to data governance as its framework does not readily scale beyond China's domestic market (Erie and Streinz 2021). Instead, global technology companies (including those based in China) will likely fragment their service offerings to account for local variation rather than adopt a company-wide standard centred on Chinese data protection (Sacks and Li 2018). Many multinational corporations have already separated their business in China from their global operations due to local factors and legal obligations such as the requirement to form a joint partnership with a Chinese company to obtain operating licences (Douglas and Feldshuh 2022).

This effect is particularly relevant for large Chinese tech firms, which have also segmented their compliance strategies between the Chinese market and abroad. For instance, Tencent and Bytedance claim in their privacy policies that they process user data differently depending on where the user registers to avoid data protection compliance risks (Tencent 2021; TikTok 2021). Personal data generated outside of China by Wechat or TikTok users is stored and processed in other jurisdictions, while data generated by Weixin or Douyin users (China's own analogues to Wechat and TikTok) remains in China. The result is that Chinese companies going abroad will not make Chinese law the de facto standard for their foreign operations.

## Transnational Effects of Chinese Law—Norms and Values

Beyond crafting a global de facto standard or binding international rules, there are other mechanisms China may use to influence global rules on data transfers. These mechanisms extend beyond hard-law instruments (such as international agreements) to soft-law mechanisms that shape international data governance norms and values (Cai 2018). Indeed, Chinese statecraft tends to prefer these avenues in certain circumstances, which makes analysing the country's influence on international law difficult for those who look primarily to formal multilateral negotiations or other forms of institutional engagement (Erie 2020).

Examples of soft-law instruments include not only technical guidelines or corporate best practices, but also a whole range of diplomatic activities that export Chinese expertise and thinking abroad (Negro 2022; Suter 2015). Such activities involve, for instance, building networks and sharing information with officials and experts from other countries, participating in international data governance forums, training programs, and peer-to-peer meetings, and working within existing governance institutions to promote Chinese interests (Benabdullah, 2020).

With respect to data transfers, there are two widely discussed sources where Chinese data ordering can shape norms and principles beyond binding international rules. First, China's localisation regime may directly or indirectly incentivise other governments to adopt similar rules or values. Second, China's participation in international fora could further entrench its norms into data governance institutions.

### *The Impact of China's Localisation Regime on Other Country's Laws and Regulations*

China's approach to data localisation may become attractive to other governments (Nanni 2020). By demonstrating the technical and regulatory possibility of exercising jurisdictional control over data, the Chinese government has begun to successfully position itself as having an effective governance model for data (Hong 2020; Mishra 2021). Indeed, if other countries began to consciously model their own laws off the PRC's system, it would signify China's exertion of transnational influence beyond corporate compliance (Erie and Streinz 2021).

Some evidence suggests that other countries have begun to model their regulatory thinking from China, but substantiating this remains a challenging task (Segal 2020). In 2022, the National Trade Estimate Report on Foreign Trade Barriers (NTE Report) found that thirty-two countries around the world have adopted some form of localisation requirements through data protection, cybersecurity, and other ICT laws (USTR 2022). Some of these laws contain structural mechanics that strongly resemble China's data governance regime. For example, Tanzania, India, Sri Lanka, Vietnam, Indonesia, Pakistan, Bangladesh, and Cambodia have all adopted or proposed rules that could limit the flow of data through Internet gateways or require organisations to receive approval from government authorities to transfer certain data abroad (USTR 2022).

However, it is difficult to determine the scope of China's influence in the formation of these regulations. On the one hand, many of these rules predate China's data localisation laws by many years (USTR 2022). For those that do not, there is little evidence that policymakers in these countries actively modelled their regulations on Chinese law, especially where legislative history of these regulations remains poorly documented (Erie and Streinz 2021).

On the other hand, the connection between localisation and national security is not unique to the PRC, which complicates pinpointing the country's role in shaping global governance trends. Indeed, Indonesia was one of the first countries to propose localisation in a general data protection law, while Russia's Federal Law No. 242-FZ from 2014 contained national security and public order justifications (Basu 2020; Bowman 2015). In each of these examples, it is difficult to disambiguate China's normative influence from other motivations. For instance, it is more likely that the regulatory trend towards data sovereignty and localisation emerged simultaneously in many countries from the same set of events, such as the Snowden Revelations in 2014 (Chander and Le 2015).

Despite this, China's effort to build regulatory capacity with other countries has and will likely increase in the future (Segal 2020). Chinese stakeholders routinely train and share knowledge with foreign officials in programs like the Baise Executive Leadership Academy, the China-ASEAN Information Port Forum, the China-Singapore Internet Forum, and the China-Africa Internet Development Cooperation Forum (Erie 2020). Such cooperation could serve as one bottom-up channel through which Chinese data sovereignty norms become more widely accepted by policymakers in key emerging jurisdictions (Cai 2018).

### *Chinese Participation in International Data Governance Institutions*

Another source of influence concerns Chinese participation in Internet governance institutions. Chinese officials and companies have actively promoted the country's data governance norms in numerous fora, including multi-stakeholder bodies like the Internet Corporation for Assigned Names and Numbers (ICANN), the World Wide Web Consortium (W3C), and the Internet Engineering Task Force (IETF) as well as multilateral organisations such as the International Telecommunication Union and the World Trade Organisation's (WTO) Joint Statement Initiative on E-Commerce (JSI). The PRC has actively emphasised its concept of cyber-sovereignty in the UN Group of Government Experts (GGE) and even helped form the Open-Ended Working Group (OEWG) with like-minded partners in 2018 after encountering US resistance to its efforts in the GGE.

China has also begun to promote data sovereignty in the institutions it created including the Shanghai Cooperation Organisation (SCO) and the Asian Infrastructure Investment Bank (AIIB In 2014, officials in Zhejiang and the CAC launched the World Internet Conference, which hosts over numerous attendees from industry, academia, and government, with participants coming from both developed and developing countries (WIC 2020). Every year, the Wuzhen Summit brings togethers data governance experts to discuss key issues and global strategic thinking around data and technology. In the past, policymakers in China promoted key cyber norms through the Summit, but in recent years the fanfare has died down. However, in 2022 Chinese officials announced that the Wuzhen Summit would become an international institution, yet details of the significance of this remain scant (Li 2022).

The PRC's position in each of these institutions varies and has changed over time. While officials have expressed scepticism for multi-stakeholder bodies, Chinese companies continue to send large delegations to many of them with government support (Negro 2020). China has articulated different goals in the ITU and the GGE than in the WTO JSI, which has largely focused on facilitating e-commerce for companies like Alibaba and JD (Gao 2022). This notwithstanding, the PRC has routinely reiterated its commitment to data sovereignty and state control of the Internet in many institutions (Erie and Streinz 2021). Indeed, the country's efforts in these bodies have strengthened perceptions that global consensus on Internet governance is fragmenting and that the trend towards stricter state control over data is irreversible (Nanni 2018).

However, many of these institutions have not placed global data transfer rules at the top of their agendas, and those that have addressed them have produced mixed results (Gao and Shaffer 2020). Despite the link between data sovereignty and localisation, China's efforts with respect to data transfers in these institutions have been inconsistent and indirect. It has committed both to promoting data flows between countries (evening signing on to the 'Data Free Flow with Trust' initiative at the G20), but also the need to uphold the sovereign rights of countries to control such data at any cost. Attempts to translate these commitments into binding agreements have advanced at a relatively slow pace, while data transfers are but one of many issues on the table (see Section II[B][1]).

One notable exception is China's Global Data Security Initiative (GDSI), which the country created on its own effort, and which directly touches upon data flows. The principles it sets forth around government access to data could help China create cross-border transfer agreements with other countries in a bilateral or plurilateral manner (Hong 2020). Yet little information beyond short readouts of these initiatives exists. Indeed, these initiatives may be driven more by a diplomatic effort to counter US and EU assertions of the lack of credibility of China's ICT products rather than a genuine attempt to craft international rules around data flows (Kak and Sacks 2021). Regardless, the initiative may help China promote its own preferred style of negotiating, especially if more countries continue to formally acknowledge it.

### Spill-Overs from Chinese Technology: Data Centres as a Pilot for Governance

The third spill-over of Chinese law and technology concerns how the export of Chinese ICT products shapes other governments' strategies with respect to data flows. In recent years, a growing body of literature has drawn connections between China's role in building infrastructure along the 'Digital Silk

Road' (DSR)—the technology component of the Belt and Road Initiative (BRI)—and its larger data governance objectives (Erie and Streinz 2021; Segal 2020). While the DSR remains more of a catchword than a comprehensive and institutionalised policy, China's technology exports continue to rise across the world, making this area an important dimension of China's growing cyber power (Creemers 2021a).

By investing, developing, and supplying the physical components and digital services that fuel the Internet, Chinese companies could exert a growing influence on global data governance. This in turn may help Beijing transmit its own values and cyber norms to other governments, particularly through the exportation of sophisticated surveillance tools, facial recognition systems, backbone ICT infrastructure, and other big data filtering devices (McKune & Ahmed 2018).

Relevant to data transfers, a notable example of these exports is the commercial spread of self-built and self-managed data centres. In recent years, Chinese policymakers and companies have made developing such centres inside and outside of the country a key priority—particularly to support cross-border data flows for Chinese companies (NDRC 2021; MIIT 2021). According to IDC's global cloud computing tracking data, Alibaba Cloud (and its unique operating system Feitian), ranks third in the global market with a share of 7.4 percent, having grown by more than 10 times in the last three years alone (Fast Technology 2022; Pandaily 2021). The company now operates in over 82 countries in 26 regions, and has long been the largest cloud provider in Asia (ICCSZ 2022). Alibaba, like other Chinese cloud providers, largely contracts with local data centres in foreign countries to expand operations (ICCSZ 2022). However, the company is beginning to commission its own data centres in foreign markets. In 2022, it launched data centres in South Korea, the Philippines, and Indonesia to accelerate its presence in APAC, with plans to expand similar centres in other key jurisdictions (ICCSZ 2022).

Additionally, state-owned enterprises (SOEs) such as China Mobile International (CMI) and China Telecom have also invested significantly in data centres abroad, particularly in the Guangdong-Hong Kong-Macao Greater Bay Area (GBA). In late 2021, CMI announced the finalisation of the Fotan Data Centre (*huotan shuju zhongxin*), which according to Hong Kong officials will form the core data centre system in the GBA (China Mobile 2021). The Fotan Data Centre forms part of CMI's global network of self-built and self-operated data centres in Singapore, London, and Frankfurt—each designed to accelerate cross-border transmission resources (China Mobile 2021). Likewise, China Telecom has also actively expanded its partnerships with data centres in numerous countries and now jointly operates more than 180 data centres across the globe (China Telecom 2020).

An emerging governance component of these data centres is managing compliance for cross-border transfers, particularly when the transmission of data implicates security assessments or data localisation requirements. This component takes on two dimensions. On the one hand, overseas data centres help Chinese companies segment their products and services between markets, particularly in circumstances where the company wishes to separate their processing activities to mitigate legal risk. Learning from this experience, Chinese data centres in third-party countries that adopt restrictions on data transfers may offer similar services for other companies.

On the other hand, data centres can become stronger vehicles for data flow security management, especially in emerging data 'free trade zones' or other digital trade programs. In China, the Ministry of Commerce's Trade in Services Plan explicitly recognises this function and directs twenty-eight test pilot zones to create dedicated data channels and supervisory models for data flows (MOFCOM 2020). Notably, the CAC and MIIT will help implement this plan through their data classification efforts by categorising and grading different types of data depending on their security risk to help organisations to map their data flows (TC260 2019). Part of this supervision model includes establishing management mechanisms that consist of data protection certifications and cross-border data flow risk assessments (MOFCOM 2020).

The goal of these measures is to facilitate the orderly flow of data by streamlining compliance with China's internal localisation requirements while promoting the use of data exchanges and other market-based mechanisms to facilitate digital trade (NDRC 2022). Policymakers in the GBA have proposed a test pilot of this cross-border management system. Numerous plans indicate the desire for Hong Kong to become a data 'customs' hub to supervise cross-border data flows, facilitate data trading in neighbouring technology centres like Shenzhen and Guangdong, and eventually serve as a bridge that connects China's data ecosystem with the rest of the world (Shenzhen 2021; Guangdong 2021). Under Chinese law, transfers of data from China to Hong Kong count as cross-border, making Hong Kong a useful test case for implementing this trial given the region's connection to the larger GBA integration process.

Under this plan, regulators envision creating a 'whitelist' mechanism to permit certain categories of data to enter Hong Kong from China and a 'negative list' to prohibit data flows that would trigger localisation under Chinese law (RCCL 2022). Such lists would green light data recipients through a CAC-led security assessment or certification that, once obtained, would last for a period of time and create legal certainty between contracting parties. Some pilots of this nature have already come online in specific sectors. For instance, the Hong Kong Monetary Authority (HKMA) and the PBOC launched a Fintech Pilot Trial Facility in 2022 to spur the cross-border transfer of financial information between major banks, third-party providers, and fintech operators (Hong Kong Monetary Authority 2022).

While details of the plan's implementation remain scant, Chinese regulators and companies may attempt to scale such 'customs' hubs beyond the GBA in partnership with other governments. Lessons learned from data free trade zones in China could also be applied in an international setting, particularly in countries that import many Chinese technology products and services (Hong 2020). Companies that operate data centres will play an important role in supervising security compliance for cross-border data transfers, both in the technical expertise they provide to regulators and their position within global data transfers.

## CONCLUSION

China's rise as a cyber power has raised implications for the future rules of cross-border data transfers. Government policy documents and other legal instruments indicate the recent coalescence of a strategy to realise two aims with respect to data transfers: (1) ensuring that data flows freely to power innovation and digital development while (2) restricting data sharing where necessary to preserve national security, regime stability, and public order. This trade-off, increasingly articulated through the concept of the *orderly flow of data*, has become a central goal of Chinese policymakers, albeit one that is scattered across ministerial departments with no unified definition.

On the global front, Chinese policymakers seek to legitimise China's cyber sovereignty while fostering data transfer rules favourable to the expansion of Chinese tech companies and China's national interests. There are notable information gaps in how this policy will be articulated and the actual pathway for realisation. While Chinese officials and companies have increased their collaboration in bilateral and multilateral fora on a range of digital governance issues, the ultimate goals of such engagement are unclear and vary depending on the forum. At a minimum, Chinese leaders have expressed their desire for non-interference in their own regulatory choices. There are examples where Chinese officials have indicated a desire to create binding international rules but also instances of collaboration that are oriented only around broad principles.

Despite the ambiguity in China's strategy, the country's current regulatory and market practices (what this chapter calls *data ordering*) have produced three spill-over effects that will likely shape the future of global data transfers in key ways. Each of these faces notable limitations that makes arriving at a conclusive answer analytically challenging. First, the solidification of China's data governance regime has shaped international expectations on data localisation and influenced how the private and public sectors understand global transfers. Yet, while China's influence on global data transfers has increased, the country's laws may have little impact beyond its domestic market.

Second, moving beyond formal rulemaking, Chinese data ordering may influence other countries by shaping expectations and norms around digital sovereignty. While there is some evidence to suggest that other countries have taken inspiration from China's approach to data governance, it is hard to isolate the degree of this influence, given that many countries have developed similar approaches simultaneously with China. Despite this, China's growing participation in international data governance institutions and its ability to attract and train foreign officials will likely strengthen its influence on future norms.

Third, Chinese companies have shaped discourse on global data flows by exporting governance through technology. Particularly, the construction of data centres in international jurisdictions contains an important governance component—cross-border security compliance as a service. Regulators have already begun experimenting with this, often partnering with SOEs and other tech companies that run data centres. Officials plan to develop a data 'customs' hub in the GBA to facilitate transfers between Hong Kong and China through streamlined compliance services. It remains unclear to what extent these hubs will be successful, but they could indicate a future trend.

## REFERENCES

Alphabeta. 2018. Micro-revolution: the new stakeholders of trade in APAC. *Alphabeta*. Available from https://alphabeta.com/wp-content/uploads/2020/06/singles-msme-report-apac.pdf [16 June 2022].

Arsene, S. 2016. Global internet governance in Chinese academic literature: rebalancing a hegemonic world order? *China Perspectives* 2, 25–35.

Arsene, S. 2018. China, internet governance, and the global public interest. In I. Sieckmann and O. Triebel, eds. *A New Responsible Power China?* (online book).

Basu, A. 2020. The retreat of the data localization brigade: India, Indonesia, and Vietnam. The Diplomat, 10 January. Available from https://thediplomat.com/2020/01/the-retreat-of-the-data-localization-brigade-india-indonesia-and-vietnam/ [17 June 2022].

Benabdallah, L. 2020. *Shaping the Future of Power: Knowledge Production and Network-Building in China-Africa Relations*. Ann Arbor: University of Michigan Press.

Bergsten, F. 2022. *The United States vs. China: the Quest for Global Economic Leadership*. Cambridge: Polity Press.

Boullenois, C. 2021. China's data strategy: creating a state-led market. *European Institute for Security Studies*. 14 October. Available from https://www.iss.europa.eu/sites/default/files/EUISSFiles/Brief_21_2021.pdf [5 May 2022].

Bowman, C. 2015. A primer on Russia's new data localization law. Proskauer. 27 August. Available from https://privacylaw.proskauer.com/2015/08/articles/data-privacy-laws/a-primer-on-russias-new-data-localization-law/ [15 June 2022].

Bradford, A. 2020. *The Brussels Effect: How the European Union Rules the World*, Oxford: Oxford University Press.

CAC. (Cyberspace Administration of China). 2017. Guanyu 'geren xinxi he zhongyao shuju chujing anquan pinggu banfa' gongkai zhengqiu yijian de tongzhi (Notice on the public consultation on the 'measures for the security evaluation of the exit of personal information and important data'), issued 11 April. Available from http://www.cac.gov.cn/2017-04/11/c_1120785691.htm [12 March 2022].

CAC. (Cyberspace Administration of China). 2019. Shuju anquan guanli banfa (zhengqiu yijian gao) (Data security management measures (draft for comment)), issued 31 May. Available from https://www.newamerica.org/cybersecurity-initiative/digichina/blog/translation-chinas-new-draft-data-security-management-measures/ [15 March 2022].

CAC. (Cyberspace Administration of China) 2022. Wangluo anquan shencha bangongshi dui zhiwang qidong wangluo anquan shencha (Cybersecurity review office launches cybersecurity review on CNKI), issued 24 June 2022. Available from http://www.cac.gov.cn/2022-06/24/c_1657686783575480.htm [25 June 2022].

CAC. (Cyberspace Administration of China) 2022a. Guojia hulian wangxi bangongshi dui didi quanqiu gufen youxian gongsi yifa zuochu wangluo anquan shencha xiangguan xingdong chufa de jueding (The Cyberspace Administration of China's decision to impose administrative penalties related to the cybersecurity review of Didi Global Co., Ltd), issued 21 July. Available from http://www.cac.gov.cn/2022-07/21/c_1660021534306352.htm [25 July 2022].

CAC (Cyberspace Administration of China). 2022b. Shuju chujing anquan pinggu banfa (Outbound data transfer security assessment measures), issued 7 July. Available from http://www.cac.gov.cn/2022-07/07/c_1658811536396503.htm [18 August 2022].

CAC (Cyberspace Administration of China). 2022c. Zhongguo guojia hulianwang xinxi bangongshi yu taiguo guojia wangluo angquan bangongshi qianshu wangluo anquang hezuo jiangjie beiwanglu (The CAC and the National Cybersecurity Office of Thailand sign a memorandum of understanding on cybersecurity cooperation), issued 5 July. Available from http://www.cac.gov.cn/2022-07/05/c_1658638472372340.htm [15 July 2022].

CAC (Cyberspace Administration of China). 2022d. Zhongguo guojia hulianwang xinxi bangongshi yu yinni guojia wangluo yu mima bu qianshu wangluo anquang hezuo xingdong jihua (The CAC and the National Network and Crytopgraphy Agency of Indonesia sign a cybersecurity cooperation action plan), issued 29 July. http://www.cac.gov.cn/2022-07/29/c_1660713333439712.htm [30 July 2022].

CAC (Cyberspace Administration of China). 2022e. Geren xinxi chujing biaozhun hetong guiding (zhenqiu yijian gao) (Standard contractual clauses for personal

information) (draft for comments)), issued 30 June. Available from http://www.cac.gov.cn/2022-06/30/c_1658205969531631.htm [15 July 2022].
Cai, Cuihong. 2018. Global cyber governance: China's contribution and approach. *China Quarterly of International Strategic Studies* 4(1), 55–76.
CAICT (China Academy of Information and Communications Technology). 2019. Da shuju bai pishu (White paper on big data), issued 15 December. Available from http://www.caict.ac.cn/english/research/whitepapers/202003/P020200327550643303469.pdf [17 May 2022].
Casalini, F. and Gonzalez, J. 2019. Trade and cross-border data flows. *Organisation for Economic Cooperation and Development*. 23 January. Available from https://www.oecd-ilibrary.org/trade/trade-and-cross-border-data-flows_b2023a47-en [15 June 2022].
CCIC (Central Cybersecurity and Informatisation Commission). 2021. 'Shisiwu' guojia xinxihua guihua (14th Five-Year Plan for Informatization), issued 28 December. Available from https://digichina.stanford.edu/work/translation-14th-five-year-plan-for-national-informatization-dec-2021/ [24 January 2022].
Central Committee and the State Council. 2020. Zhonggong zhongyang guowuyuan guanyu goujian geng jia wanshan de yaosu shichanghua peizhi tizhi jizhi de yijian (Opinions of the central committee and the state council on building a more perfect market-based allocation system and mechanism for factors), issued 9 April. Available from http://www.gov.cn/zhengce/2020-04/09/content_5500622.htm [16 January 2022].
Chander, A and Le, U. 2015. Data nationalism, *Emory Law Journal* 64(3), 677–739.
Chander, A. and Sun, H. 2021. Sovereignty 2.0. *University of Hong Kong Faculty of Law*
*Research Paper No. 2021/041*. 8 September. Available from https://papers.ssrn.com/sol3/papers.cfm?abstract_id=3904949 [January 20 2022].
China Mobile. 2021. Zhongguo yidong yue gangao da wanou xianggang huotan jushu zhongxin zhengshi donggong, zhuli dazao xianggang guoji chuangxin keji zhongxin (China Mobile's Guangdong-Hong Kong- Macao Greater Bay Area Fotan Data Center officially started construction, helping build Hong Kong's international innovation and technology center), *zhongguo yidong* (China Mobile), 8 December. Available from https://www.10086.cn/aboutus/news/groupnews/index_detail_40650.html [16 February 2022].
China Telecom Americas. 2022. Global data center map. *China Telecom Americas* Available From https://www.ctamericas.com/global-data-center-map/#:~:text=China%20Telecom%20operates%20450%2B%20on,data%20centers%20across%20the%20globe [14 June 2022].
Clarke, D. 2020. Order and law in China. *GWU Legal Studies Research Papers*, no. 2020–52. 25 August. Available from https://papers.ssrn.com/sol3/papers.cfm?abstract_id=3682794 [22 August 2022].
Creemers, R. 2020. China's conception of cyber sovereignty: rhetoric and realization. In D. Broeders & B. van den Berg, ed., *Digital Technologies and Global Politics*. Lanham: Rowman & Littlefield, 107–142.

Creemers, R. 2021. China's emerging data protection framework (forthcoming). Available from https://papers.ssrn.com/sol3/papers.cfm?abstract_id=3964684.

Creemers, R. 2021a. The Digital Silk Road: perspectives from affected countries. *Leiden Asia Centre.* July. Available from https://leidenasiacentre.nl/report-the-digital-silk-road-perspectives-from-affected-countries/ [14 April 2022].

Douglas, A. and Feldshuh, H. 2022. How American companies are approaching China's data, privacy, and cybersecurity regimes. *U.S.-China Business Council.* April. Available from https://www.uschina.org/reports/how-american-companies-are-approaching-china%E2%80%99s-data-privacy-and-cybersecurity-regimes [12 May 2022].

Erie, M. 2020. Chinese law and development. *Harvard International Law Journal* 62(1).

Erie, M. and Streinz, T. 2021. The Beijing effect: China's 'Digital Silk Road' as transnational data governance. *New York University Journal of International Law and Politics* 54(1), 1–92.

Fast technology. 2022. Er ling er yi nian quanqiu yun jisuan paiming: zhongguo aliyun jishen qian wu (Global cloud computing rankings 2021: China's Alibaba cloud among the top five). Fenghuang wang keji (Phoenix Technology Net) 28 April. Available from https://tech.ifeng.com/c/8FaEw9l7k1E [14 June 2022].

Gao, H. 2018. Digital or trade? The contrasting approaches of China and the US to digital trade. *Journal of International Economic Law* 21(2), 297–321.

Gao, H. and Shaffer, G. 2020. A new Chinese economic order? *Journal of International Economic Law* 23(3), 607–35.

Gao, H. 2022. Data sovereignty and trade agreements: three digital kingdoms. *Hinrich Foundation* 18 January. Available from https://www.hinrichfoundation.com/research/article/digital/data-sovereignty-trade-agreements-digital-kingdoms / [17 August, 2022].

Gu, S. and Lundvall, B. 2006. China's innovation system and the move towards harmonious growth and endogenous innovation. Druid Working Papers 06/07. Available from https://ideas.repec.org/p/aal/abbswp/06-07.html [17 August 2022].

Guangdong. 2021. Guangdong sheng shuju yaosu shichanghua peizhi huang xingdong fan'an (Action plan for the reform of market-oriented allocation of data elements in Guangdong province), issued 11 July 2021. Available from http://www.gd.gov.cn/xxts/content/post_3342648.html [3 June 2022].

Hong Kong Monetary Authority. 2022. Launch of Greater Bay Area fintech pilot trial facility, issued 18 February. Available from https://www.hkma.gov.hk/media/eng/doc/key-information/guidelines-and-circular/2022/20220218e2.pdf [17 May 2022].

Hong, Yanqing. 2017. Ping 'wangluo anquan fa' dui shuju anquan baohu zhi de yu shi (On the gain and loss of the Cybersecurity Law of China on data protection). *China Academic Journal Electronic Publishing House*. 7 November. Available from http://library.ttcdw.com/uploadfiles/zk/1507792951.pdf [15 July 2022].

Hong, Yanqing. 2020. Chinese opinion to speed up transborder data flow under the BRI. *China Law*, 94–100.

Huang, Jie. 2020. Applicable law to transnational personal data: trends and dynamics. *German Law Journal* 21(6), 1283–1308.

Hummel, P. et al. 2021. Data sovereignty: a review. *Big Data and Society* 1: 1–17.

ICCSZ. 2022. Ali yun hanguo shuju zhongxin zhenshi qiyong jiasu yatai shichang buju (Alibaba cloud South Korea data centre officially opened to accelerate Asia-Pacific market layout), *xunshi guangtong xunwang* (ICCSZ), 31 March. Available from http://www.iccsz.com/site/cn/News/2022/03/31/20220331023932348574.htm [15 June 2022].

Kak. A and Sacks, S. 2021. Shifting narratives and emergent trends in data-governance policy: developments in China, India, and the EU. *Yale Law School Paul Tsai China Center*. 8 August. Available from https://law.yale.edu/sites/default/files/area/center/china/document/shifting_narratives.pdf [17 August 2022].

Kong, T. and Tong, S. 2021. China's Comprehensive and Progressive Agreement for Trans-Pacific Partnership application. *East Asian Institute Commentary* 37. 4 October. Available from https://research.nus.edu.sg/eai/wp-content/uploads/sites/2/2021/10/EAIC-37-20211004.pdf [15 August 2022].

Kuner, C. 2011. Regulation of transborder data flows under data protection and privacy law, *OECD Digital Economy Papers* 187.

Li, Jiaxing. 2022. China's World Internet Conference goes 'international' as Beijing seeks to promote its own vision of global cyberspace. South China Morning Post. 13 July 2022. Available from https://www.scmp.com/tech/big-tech/article/3185151/chinas-world-internet-conference-goes-international-beijing-seeks [13 July 2022].

Liu, Lizhi. 2021. The rise of data politics: digital China and the world. *Studies in Comparative International Development* 56(3), 45–67.

Lu, Chuanying. 2014. Zhuquan gainian de yanjin jiqi zai wangluo shidai mianlin de tiaozhan (Evolution of the concept of sovereignty in the challenges of the internet age). *Guoji guanxi yanjiu* (International Relations Studies 1(3), 75–77.

Lu, Chuanying. 2016. China's emerging cyberspace strategy. The Diplomat. 24 May. Available from https://thediplomat.com/2016/05/chinas-emerging-cyberspace-strategy/ [14 June 2022].

Lu, Xiaomeng. 2020. Is China changing its thinking on data localization?. The Diplomat. 4 June. Available from https://thediplomat.com/2020/06/is-china-changing-its-thinking-on-data-localization/ [15 July 2022].

McAuliffe, Z. 2022. Tiktok is the most downloaded app worldwide in 2022 so far, analyst says. *CNET*. 14 April. Available from https://www.cnet.com/news/social-media/tiktok-is-the-most-downloaded-app-worldwide-in-2022-so-far-analyst-says/ [25 August 2022].

McKune, S. and Ahmed, S. 2018. The contestation and shaping of cyber norms through China's internet sovereignty agenda. *International Journal of Communication* 12, 3835–55.

MFA (Ministry of Foreign Affairs). 2015. Remarks by H.E. Xi Jinping President of the People's Republic of China at the opening ceremony of the second World Internet Conference, issued 16 December. Available from https://www.fmprc.gov.cn/eng/wjdt_665385/zyjh_665391/201512/t20151224_678467.html [13 January 2022].

MFA (Ministry of Foreign Affairs). 2017. International strategy of cooperation on cyberspace, issued 3 January. Available from https://www.fmprc.gov.cn/mfa_eng/wjb_663304/zzjg_663340/jks_665232/kjlc_665236/qtwt_665250/201703/t20170301_599869.html#:~:text=Cyberspace%20is%20the%20common%20space,of%20shared%20future%20in%20cyberspace [23 August 2022].

MFA (Ministry of Foreign Affairs). 2020. Global Initiative on Data Security, issued 8 September. Available from https://www.mfa.gov.cn/ce/ceus//eng/zgyw/t1812951.htm [15 January 2022].

MIIT (Ministry of Industry and Information Technology). 2021. Xinxing shuju zhongshu fazhan sannian xingdong jihua (Three-year action plan for the development of new data centres), issued 4 July. Available from http://www.gov.cn/zhengce/zhengceku/2021-07/14/content_5624964.htm [15 March 2022].

Mishra, N. 2019. Building bridges: international trade law, internet governance, and the regulation of data flows. *Vanderbilt Law Review* 52, 463–509.

MLex 2022. Singapore proposes mutual recognition of data-protection certification scheme with China. Available from https://mlexmarketinsight.com/news/insight/singapore-proposes-mutual-recognition-of-data-protection-certification-scheme-with-china [16 November 2022].

MOFCOM (Ministry of Commerce). 2020. Quanmian shenhua fuwu maoyi chuangxin fazhan shidian zongti fan'an (Overall plan for comprehensively deepening the pilot program for innovative development of trade in services), issued 14 August. Available from http://images.mofcom.gov.cn/fms/202008/20200814092010526.pdf [20 February 2022].

Mueller, M. 2017. *Will the Internet Fragment?: Sovereignty, Globalization, and Cyberspace*, New York: Wiley.

Muhlhahn, K. 2019. *Making China Modern: From the Great Qing to Xi Jinping* Cambridge: Harvard University Press.

Nanni, R. 2020. Rising China and the global internet: assessing China's challenge to the global internet governance system and the international liberal order. *Giganet* Available from https://www.giga-net.org/2020symposiumPaper/Nanni.pdf?_t=1602675821 [5 September 2022].

Negro, G. 2020. A history of Chinese global internet governance and its relations with ITU and ICANN. *Chinese Journal of Communication* 13(1), 104–21.

NDRC (National Development and Reform Commission). 2021. Quanguo yitihua da shuju zhongxin xietong chuangxin tixi suanli shuniu shishe fan'an (Implementation plan for the computing power hub of the collaborative innovation system of the national integrated big data center), issued 24 May. Available from http://www.gov.cn/zhengce/zhengceku/2021-05/26/content_5612405.htm [14 June 2022].

NDRC (National Development and Reform Commission). 2022. Guanyu dui 'shuju jichu zhidu guandian' zhengji yijian de gonggao (Announcement on the solicitation of comments for the 'views on the data foundation system'), issued 21 March. Available from https://hd.ndrc.gov.cn/yjzx/yjzx_add.jsp?SiteId=378 [14 June 2022].

NPC (National People's Congress). 2021. Zhonghua renmin gongheguo geren xinxi baohu fa (Personal Information Protection Law). Available from https://digichina

.stanford.edu/work/translation-personal-information-protection-law-of-the-peoples-republic-of-china-effective-nov-1-2021/.

NPC (National People's Congress). 2021a. Zhonghua renmin gongheguo shuju anquan fa (Data Security Law). Available from https://digichina.stanford.edu/work/translation-data-security-law-of-the-peoples-republic-of-china/

Pandaily. 2022. Alibaba cloud ranks third in global cloud computing market in 2021. *Pandaily* 28 April. Available from https://pandaily.com/alibaba-cloud-ranks-third-in-global-cloud-computing-market-in-2021/#:~:text=Cloud%27s%20market%20share%20took%20up,to%20enter%20the%20top%20three [14 June 2022].

Park, Chaeri. 2022. Knowledge Base: China's 'Global Data Security Initiative.' *DigiChina*. 31 March. Available from https://digichina.stanford.edu/work/knowledge-base-chinas-global-data-security-initiative/ [24 August 2022].

Pongratz, S. 2020. Key takeaways—the telecom equipment market 1H20. Dell'Oro Group. 7 September. Available from https://www.delloro.com/key-takeaways-the-telecom-equipment-market-1h20/ [August 18 2022].

Qiushi. 2021. Bu duan zuoqiang zuoyou zuoda wo guo shuzi jingji (Continuously make our country's digital economy stronger, better, and bigger). Speech from Xi Jinping. 18 October. Available from http://www.qstheory.cn/dukan/qs/2022-01/15/c_1128261632.htm [17 May 2022].

RCCL. 2019. Legal research project: proposal for Hong Kong to be a data center hub for the Greater Bay Area and China. *Centre for Chinese and Comparataive Law*, 20 December. Available from https://beltandroad.hktdc.com/en/insights/legal-research-project-proposal-hong-kong-be-data-center-hub-greater-bay-area-and-china [6 April 2022].

Sacks, S. and Li, M. 2018. How Chinese cybersecurity standards impact doing business in China. *Center for Strategic and International Studies*. 2 August. Available from https://www.csis.org/analysis/how-chinese-cybersecurity-standards-impact-doing-business-china [15 June 2022].

Segal, A. 2020. China's alternative cyber governance regime. *U.S. China Economic Security Review Commission* 13 March. Available from http://www.uscc.gov/sites/default/files/testimonies/March%2013%20Hearing_Panel%203_Adam%20Segal%20CFR.pdf [21 June 2022].

SCIO (State Council Information Office). 2010. White paper on the internet in China, issued 8 June. Available from http://www.chinadaily.com.cn/china/2010-06/08/content_9950198.htm [4 January 2022].

Shenzhen. 2022. Shenzhen jiang jian xinxing shuju maoyi xinxihua pingtai peiyu yao 5 jia zhiming kuajing shuju shang (Shenzhen will build a new data transaction information platform to cultivate roughly five well-known cross-border data providers), issued 16 May. Available from http://www.sz.gov.cn/cn/xxgk/zfxxgj/zwdt/content/post_9783306.html [16 June 2022].

State Council. 2015. Cujin da shuju fazhan xingdong wangyao (Action plan for promoting big data development), issued 31 August. Available from http://www.gov.cn/zhengce/content/2015-09/05/content_10137.htm [14 May].

State Council. 2016. Shisanwu goujia xinxi guihua (Thirteenth Five-Year Plan for National Informatization), issued 15 December. Available from http://www.gov.cn/zhengce/content/2016-12/27/content_5153411.htm [24 March 2022].

Suter, D. 2015. The Shanghai Cooperation Organisation of a Chinese practice of international law. *Zurcher Studien zum offentlichen Recht* 232.

TC260 (National Information Security Standardization Technical Committee). 2022. Wangluo anquan biaozhun shixian zhinan, yi geren xinxi kuajing chuli huodong renzheng jishu guifan (Technical specifications for certification of cross-border processing activities of personal information), issued 24 June. Available from https://www.tc260.org.cn/front/postDetail.html?id=20220429181520 [25 August 2022].

TC260 (National Information Security Standardization Technical Committee). 2019. Wangluo anquan biaozhun shixian zhinan—wangluo shuju fenlei fenji zhiyin (Network security standard practice guidelines—guidelines for categorization and grading of network data), issued 31 December. Available from https://www.tc260.org.cn/front/postDetail.html?id=20211231160823 [18 August 2022].

Tencent. 2021. Privacy Policy. *Tencent*. Available from https://www.tencent.com/en-us/privacy-policy.html [25 May 2022].

TikTok. 2021. Privacy Poicy. *TikTok*. 5 October. Available from
https://www.tiktok.com/legal/privacy-policy-eea?lang=en [25 May 2022].

UNCTAD (United Nations Conference on Trade and Development). 2021. Digital economy report. 2021. *United Nations*. 29 September. Available from https://unctad.org/system/files/official-document/der2021_en.pdf [17 June 2022].

USTR (United States Trade Representative). 2022. National trade estimate report on foreign trade barriers, issued 31 March. Available from https://ustr.gov/sites/default/files/2022%20National%20Trade%20Estimate%20Report%20on%20Foreign%20Trade%20Barriers.pdf [16 April 2022].

Voss, G. 2020. Cross-border data flows, the GDPR, and data governance. *Washington International Law Journal* 29(3), 485–532.

Weng, Guomin and Song, Li. 2020. Shuju kuajing chuanshu de falu guiding (Legal regulation of cross-border data transfer). *Zhejiang daxue xuebao* (*Journal of Zhejiang University*) 50(2), 38–53.

WIC (World Internet Conference). 2020. Overview of WIC. *World Internet Conference*, 15 October. Available from https://www.wuzhenwic.org/2020-10/15/c_547699.htm [23 June 2022].

Williams, B. and Sutherland, M. 2021. China and Taiwan both seek to join the CPTPP. *Congressional Research Service*. 24 September. Available from https://crsreports.congress.gov/product/pdf/IN/IN11760 [22 August 2022].

Wu, Xuan. 2021. Shuju zhuquan shiye xia geren xinxi kuajing guize de jiangou (The construction of personal information cross-border rules from the perspective of data sovereignty), *Tsinghua faxue* (Tsinghua University Law Journal) 15(2), 74–91.

Yakoleva, S. and Iron, K. 2020. Pitching trade against privacy: reconciling EU governance of personal data flows with external trade. *International Data Privacy Law* 10(3), 201–21.

Yang, Rongjun. 2014. Lun wangluo kongjian zhili guoji hezuo mianlin de nanti jiqi yingdui celue (On problems and strategies of international cooperation in cyberspace governance), *guangxi shifan daxue xuebao* (Guanxi Normal University Journal of Social Studies) 13(3), 79.

Ye, Kairu. 2020. Shuju kuajing liudong guizhi zhong de 'changbi guanxia'(Long-arm jurisdiction in the rules of cross-border data flows), *faxue pinglun* (Legal Review) 1: 106–17.

Zanfir-Fortuna, G. 2021. Dispatch from the global privacy assembly: the brave new world of international data transfers. Future of Privacy Forum. 10 November. Available from https://fpf.org/blog/dispatch-from-the-global-privacy-assembly-the-brave-new-world-of-international-data-transfers/ [18 August 2022].

Zhang, Jinping. 2016. Shuju kuajing chuanshu de gouji guize yu zhonguo guizhi fanying (International rules for cross-border data transfer and China's regulatory reactions), *zhengzhi yu falu* (Politics and Law).

Zhang, Longmei and Chen, S. 2019. China's digital economy: opportunities and risks. International Monetary Fund, Working Paper 19/16. Available from https://www.imf.org/~/media/Files/Publications/WP/2019/wp1916.ashx [August 18 2022].

Zheng, Weiwei. 2020. Comparative study on the legal regulation of a cross-border flow of personal data and its inspiration to China. *Frontiers of Law in China* 15(3), 280–312.

Zheng, Wei. 2022. Wei shuji maoyi guoji guize zhiding gongxian zhongguo de zhihui (Contributing Chinese wisdom to the formulation of international rules for digital trade). *Ministry of Commerce*. 10 February. Available from http://tradeinservices.mofcom.gov.cn/article/yanjiu/pinglun/202202/130233.html [24 August 2022].

*Chapter 7*

# China and Global Internet Governance

## *ITU, ICANN, and the World Internet Conference*

Gianluigi Negro

In line with the rationale of the book, this chapter explores the engagement of the Chinese government into the Internet governance discussion. Previous studies have already highlighted that, during the last two decades, China has been shifting its role into the internet governance discussion from norm taker to a norm maker one (Galloway, 2015; Negro, forthcoming). This contribution shows how China has been increasing its presence into two of the most relevant international organisations in the field of the internet governance: the International Telecommunication Union and the Internet Corporation for Assigned Numbers (ICANN). This chapter also focuses on the role of the World Internet Conference (also known as the Wuzhen Summit), a Chinese initiative launched by the Chinese government to promote an alternative vision of global Internet governance.

The analysis of these three cases studies is based on official documents and specialised Chinese academic journals and magazines and it supports three main arguments. First, the Chinese idea of Internet governance cannot be limited to the role of the government, whereas it involves a variety of stakeholders that, at least form a historic perspective, have not been always fully matching the state's interests (Shen, 2016). Views of policymakers, scholars and businesspersons are not always convergent with official statements. This trend shows a degree of inconsistency in the Chinese narrative.

The second argument raised from this contribution challenges the idea according to which China's contribution to the Internet governance discussion is limited to a dichotomy between a multi-stakeholder model, closer to US values and the *status quo* of the Internet governance, and a multilateralism, a vision often promoted by the Chinese official narrative but that does not fully reflect the complex vision of Chinese Internet governance, which de facto match a series of multi stakeholder principles. In general terms, the second argument challenges the idea according to which China is contributing to the fragmentation of the Internet (Guan, 2019; Lindsay, Cheung, and Reveron, 2015) or paradigms as the Internet Yalta (Klimburg, 2013) and the Digital Cold War (O'Connor, 2014).

The third argument of this this contribution highlights the China's ambition to play a more pivotal role into the Internet governance discussion, as it is demonstrated by the creation of the World Internet Conference, an original space to foster and coordinate an alternative vision of the global Internet governance.

## CHINA AND ITU

There are at least two reasons to support the relevance of ITU for the Chinese vision on the global Internet. First, in the field of telecommunications, it is an international organisation that is not structured along multistakeholder lines (Raymond and DeNardis 2013). Indeed, although it includes international organisations, NGOs, firms and academic institutions in its decision making processes, the main decision making powers on the regulation of international telecommunication are reserved to the ITU's member states. The preponderance of the member state's contributions and their sovereign rights to determinate Internet policies and regulation form the core of the "multilateral" model of Internet governance (Bauer and Dutton 2015). In other words, the ITU approach is in line with the China's idea of "Internet sovereignty" (*hulianwang zhuquan*) defined on December 2016 by the National Cyberspace Security Strategy as states' right to "prevent curb and publish the online dissemination of harmful information endangering national security and interests, and to safeguard order in cyberspace" (CAC 2016).

The second strategic reason behind China's interest and growing presence at the ITU is justified by the role played by a particular section called ITU-T and aimed at coordinating standards for telecommunications and ICTs such as cybersecurity, machine learning and video compression. Participating in standards-setting at the global level provides several political and economic advantages as Tim Rühlig's chapter in this volume discusses in depth. Previous studies argue that standards can seizure the definition of a particular technology trajectory, the direction and, to some extent, the rate at which future technology progress develops (Suttemeier and Yao 2008).

ITU and China relations are important also from a historical and symbolic perspective. ITU is the first international organisation in the field of telecommunications. In 1994, it sent a delegation to China to support the country in creating its first Internet telecommunication infrastructure, it also supported China to develop a *know how* among Chinese engineers in the field of Information Communication Technologies through a series of ITU sponsor seminars and workshops. At the time of writing, Zhao Houlin was serving his second term as ITU Secretary General. Furthermore, during the last two decades, China has increased its presence also in specific technical committees and secretariats. The presence of China in ITU is not limited to Chinese officials but includes also important private sector actors such as Huawei. The Shenzhen based company, played a crucial role in the proposal of a the "New-IP" project, a choice that can be read as strategic because it sees a direct engagement of a Chinese company to support a state-centric approach in the standard setting process. At this concern, it should be also noted that at the December 2020 plenary session of ITU-T Study Group 11 and 15 it was decided to not accept "New IP" repeated questions as new work items and to stop discussing "New IP" at least until the World Telecommunication Standardisation Assembly that took in March 2022. However, since that decision documents that support "New IP" proposal continued to appear in forms of new proposals in different study groups at ITU-T (Drolet 2022). Huawei experience apart, the growing presence at Geneva headquarters of Chinese delegates and sector actors in the last years suggests more confidence to influence the decision making process at ITU compared to other standards developing organisations such as IEFT and ICANN.

Beside private sector actors, important Chinese contributions to ITU also come from the academic sector, including Tsinghua University, Wuhan University and the Beijing University of Posts and Telecommunications. The rising role of the Chinese academic sector at ITU is further illustrated by a memorandum of understanding co-signed by ITU and Tsinghua University in January 2019 aimed at launching the academic journal *ICT Discoveries*. This is intended to promote academic debate on the latest ITCs technical developments and their policy, regulatory, economic, social, and legal dimensions.

In more general terms, at the time of writing the number of Chinese delegates in the telecommunications section of ITU (ITU-T) is second only to the United States, and ahead of Japan (ITU, List of Sector Members 2022). However, the presence of delegates in the ITU-T does not necessarily mean a concrete influence. Indeed, within the working groups of ITU-T the role played by chairs and vice chairs has more influence than delegates. At the present stage, China has a chair position only in the SG16, a study group that focuses its activities on multimedia coding, system, and application (Negro, forthcoming). Another important position is represented by the vice-presidency in SG15, this working group actions are on networks and infrastructure for transport, access, and home. The Chinese presence in these working groups does not imply the adoption of technical standards. Nevertheless, the contribution provided by Chinese companies, research bodies and government delegates can impact the parameters and the negotiations for the standardisation process.

China's relationship with the ITU has deep historical roots. Having joined officially in 1920, China sent its first delegation to the 1932 Plenipotentiary Conference in Madrid. The Nationalist Chinese government also signed the ITU Convention. In 1947, at the ITU Plenipotentiary in Atlantic City, China was elected for the first time to the ITU Executive Council. Furthermore, in the same year, Chinese was adopted as an official ITU language. After the establishment of the People's Republic of China, in line with the rest of the UN system, the ITU continued to recognise the Nationalist Chinese government in Taipei, until the changeover to Beijing in 1972. From that moment Taiwan and the territories controlled by the Republic of China (ROC), has a country code and are listed as "Taiwan, China."

In terms of Internet governance, ITU played a very important role both in terms of *know-how* and infrastructure sending a delegation to China in 1994, the year China was officially connected to the World Wide Web (Negro 2020). The results of this cooperation are summed up on the report of "China-ITU Seminar for Strategy for Telecommunication Development" held in Beijing between the 27th and the 30th of June 1994. From that year on ITU supported China providing scientific funding and consulting in the development of the Chinese Internet and telecommunications infrastructure. The cooperation between China and ITU intensified in 2005 when the ITU-United Nations supported initiative World Summit of Information Society (WSIS) officially issued a definition of "Internet governance," presented as "the development and applications by governments, the private sector and civil society, in their respective roles, of shared principles, norms, rules and decision making, procedures, and programs that shape the evolution and use of the Internet." The World Summit of Information Society was an ITU action divided into two phases (2003 in Geneva and 2005 in Tunis), one of the most important goals was to modify ICANN structure. Indeed, The Tunis Agenda for the Information Society, the official document produced during the second phase required to "enable governments, on an equal footing, to carry out their roles and responsibilities, in international public policy issues pertaining to the Internet, but not in the day-to-day technical and operational matters, that do not impact on international public policy issues (WSIS, 2005, art. 69). The WSIS initiative did not reach the expected results, its output was limited to the creation of the Internet Governance Forum (IGF), an annual event based on the idea of the multistakeholder approach, but with an only deliberative power. Furthermore, after two years from the WSIS failure, in 2007, Zhao Houlin, at that time president of the ITU Telecommunication Standardisation Bureau, published a research article on the academic journal *Information Polity* suggesting the break away from ICANN centralised administration of Internet domain names to move towards the decentralised administration that characterises other telecommunications naming and addressing resources (in particular, telephone numbers) (Zhao 2007, see also Negro, 2022).

The two WSIS phases were seen strategic for China because they represented a chance to shift the management of the Internet resources from ICANN to ITU, an international organisation with a closer vision to its idea of internet sovereignty but it was also important for the ITU to gain support from China and other developing countries to take over ICANN's role in its management of Internet resources.

This process represented a shift from earlier definitions of Internet governance, focusing on the global technical management of the Internet's core resources: domain names, IP addresses, Internet protocols and the root server system (Kleinwächter 2004). All these issues were already regulated by ICANN, which it was established in 1998 thanks to the support of the US Clinton-Gore administration, with the aim to support neoliberalist values as well as a network of interests among the technical community, the US government, intellectual property rights holders and other agents form the private sector (Mathison 2009). There are at least two reasons that could contextualise the neoliberalist approach. The first one is the publication of the *White Paper on Internet Governance* issued by the Clinton administration, a document that influenced the creation of ICANN and focused on the importance of individual freedom and on the protection of private property and a commitment to economic laissez-faire (Harvey 2005). Furthermore, as Chenou aptly notes, the Clinton administration issued another document a few months before the publication of the *White paper* on which the authors suggest an active role in the creation of new self-reliant markets (Chenou 2014). In other words, the main goal to create a neoliberal legal framework of the Internet was the creation of a market for domain names where the institutions and rules are designed to ensure the straightforward functioning of the market. The role of the institutions is to ensure the stability of the network infrastructure and to safeguard private intellectual property (ICANN & DoC, art II). According to this neoliberal approach the Internet needs to be regulated by an individualised and market-based competition, which is considered the most appropriate expression of governance compared to other forms of organisation.

The second reason that justifies the neoliberalist approach is the structure of ICANN itself, which *de facto* does not support intergovernmental forms of governance. Even its Governmental Advisory Committee appoints a non-voting liaison to the ICANN Board. Whereas the ICANN board runs the market-enabling institutional role supporting a private, transnational, and not for profit cooperation.

The institution of ICANN reflected the phenomenon of "Internet exceptionalism" (Chenou 2014), justifying the decision to bypass the ITU. Indeed, since its foundation, ICANN has overseen allocating top-level domains (ccTLD). Finally, ICANN's nature as a private company created a new market for Internet domain names based on the self-organisation of the Internet community, without the interference of government (Bygrave & Bing 2009).

To achieve its goal, China played an active role in the activities of the Working Group on Internet Governance (WGIG), a platform created to collect ideas and proposal, which eventually published a series of reports on Internet governance during the two phases of the WSIS (Shen 2016). The Chinese presence into WGIG did not only include Chinese official but also private corporations like ZTE and Huawei.

During the two WSIS phases, Chinese delegations officially lamented how "Internet governance was monopolised by one state, one corporation or a handful of private corporations" (Sha 2003). Chinese authorities considered the ITU and UN, as well as the events they sponsored, as the appropriate venues to support voices from the Global South. The Chinese Ministry of Information Industry stated at the first WSIS meeting that "developing countries, through their own efforts, explore development modes of information society that suit their own national conditions, and China will work unremittingly towards this end" (2003). It is interesting to note that already during WGIG activities and the first WSIS phase, China adopted a pivotal role for the Global South discussion in the field of the Internet governance.

If, sometimes, the Chinese approach to Internet governance seems inconsistent and sometimes even contradictory, its engagement with ITU and WGIG in particular, can be considered rational. Indeed, the Chinese participation to WGIG activities reflects both an occasion to the Chinese government to express its own view on Internet governance but also a moment to obtain credibility within ITU. It also foreshadowed several key points that would gain greater priority in the decades later. At a 2004 WGIG meeting, Hu Qiheng, then advisor for the Science and Technology Commission of the Ministry of Information Industry and vice president of the Chinese Academy of Social Sciences, stated that "Internet governance and the administration of the domestic Internet falls within the sovereignty of each country," claiming the interests of a state and its people are best represented by governments, and that private sector and civil society actors cannot do so (2004; see Negro,

2020). All in all, during this stage, China clearly expressed its criticism towards the *status quo* of the Internet governance. China kept on investing its efforts on United Nations and ITU initiatives also after the two WSIS phases.

That said, ICANN is still one of the main international organisations in the field of the Internet governance. This not necessarily means that the China strategy has been unsuccessful. China's contribution to the ITU has been beneficial to the international organisation at least to challenge the role of ICANN and to start an international discussion on Internet governance. Furthermore, ITU is now one of the most important international organisations where China can propose its own vision of Internet governance outside its borders establishing forms of cooperation with other countries. The Chinese presence into ITU technical sectors study groups is important because, although they could not directly impact the operations of the Internet, they are relevant for setting international standards in the global infrastructure of ICTs (ITU-T), for managing radio systems (including satellite ownership and spectrum allocation) (ITU-R) and for closing the digital divide providing technical and capacity service for developing countries.

## CHINA AND ICANN

The relationship between China and ICANN has not been particularly linear and has evolved in four distinct stages. First, between the late 1990s and the beginning of the new millennium. the China–ICANN relation was mainly formal and inconsistent. Subsequently, during the 2000s, a series of conflicts and contrasts led to a clear divergence of visions on global Internet governance. Third, from 2009 onwards, Chinese official delegations reintegrated with ICANN. Finally, from 2016 on, after the IANA transition, which severed ICANN's formal links with the US Department of Commerce, China's reaction was largely positive and led to an ongoing discussion on how China can enhance its position in the current ICANN arrangements.

The first official encounter between ICANN and a (nongovernmental) Chinese delegation took place in 1999, five years after China officially gained accessed to the Internet, Tsinghua Professor Wu Jianping was elected a member of ICANN's Address Supporting Organisation. The same year, the deputy bureau director of the Ministry of Information Industry (MII)[1] Chen Yin, represented China at the meeting of the ICANN Governmental Advisory Committee (GAC), a body with a limited power in the Internet domain politics.

Divergences emerged very soon, for two reasons. A first problem was the formal acknowledgment of Taiwan as an independent country in the GAC (Mackinnon, 2009), an agreement was reached in 2000 after ICANN also agreed to refer to the island as "Chinese Taipei" (ibid.). The second reason involved a case where a Virginia court ordered the Chinese company Maya to relinquish ownership of the CNNnews.com after a domain name squatting complaint from CNN, even though Maya had legally acquired it from a China-based domain name registrar. In response, China suspended official delegations to GAC meetings from 2001 to 2009. However, this decision did not compromise other engagements of Chinese individuals and Internet operators (Shen 2016). Qian Hualin, deputy director of the China Internet Network Information Center (CNNIC), China's domain registry, served as ICANN Board of Director from 2003 to 2006. Moreover, an ICANN meeting took place in in Shanghai in 2002 with full support of the Internet Society of China and CNNIC. It is interesting to note that although China government officials did not take part to GAC activities, China played an active role through the engagement to ICANN activities through the Internet Society of China, one of its most representative nongovernmental organisations with more of four hundred members in the field of industry and academia. Even more interesting, this support took place at the same moment ICANN formalised the creation of a Support Organisation to represent country code interests in ICANN replacing the Domain Name Supporting Organisation (DNSO)[2] (ICANN, 2002). Due to these circumstances, it is possible to argue that the Chinese engagement with ICANN remained ambiguous.

In 2009, China resumed participation in the GAC, while ICANN implemented two important measures. The first measure was to reform the domain space by allowing any established entity located everywhere in the world to operate a new TLD registry (Zhu, 2012). This operation opened *de facto* the domain name market with a global bid for the creation and management of new general top-level domains (gTLDs) (Arsene, 2015) leading to an expansion for the market and a reshuffle of the registrar and registry industry. Chinese institutions such as MIIT and CNNIC supported Chinese companies to occupy the new domain landscape limiting the entrance of foreign registries such as Verisign and Neustar and transnational registrars such as GoDaddy and Tucows. The Chinese rush to occupy a reformed domain market can be justified not only by a protectionist move by the Chinese institutions but also by the spending power of Chinese registrars and registries. Obviously, this phenomenon was beneficial to ICANN. This historical change reveals a contradiction in the Chinese Internet governance strategy. On the one hand, Chinese officials during the two WSIS phases, lamented the unilateral US management (ICANN in particular) in the field of Internet resources; on the other hand, with huge and fast investments into the new domain names market China privileged the private interests of its registrars and registries supporting the ICANN market driven approach. The economic impact of this first measure was clear in 2016, that year China covered 54 percent of global domain names with new gTLD extensions, with an increase of 400 percent between 2013 and 2015 (CNBC, 2016).

The second measure was that ICANN created "internationalised" TLDs in non-Roman letter scripts. This new scenario opened a new market for Internet addresses, amongst others in Chinese characters, with significant economic potential.

It also should be noted that, even before the IANA transition, China strengthened its cooperation with ICANN. In 2013, the 46th ICANN meeting took place in Beijing, then the most-attended event in terms in ICANN's history. At that occasion, the GAC issued the Beijing Communiqué, a document outlining a series of "safeguards" on top-level domains and suggesting "a public requirement" for the approval of new "exclusive registry access" gTLDs (ICANN, GAC, 2013). Also, in the same occasion, ICANN opened its first Engagement Centre in Beijing in order to strengthen collaboration with Chinese authorities (ICANN, 2013). Interestingly, the announcement was made by Hu Qiheng, the Chinese scholar who during a WGIG event in 2004 had criticised ICANN for its lack of transparency and attention to the Global South. Subsequently, the MIIT's China Academy of Telecommunications Research concluded a Memorandum of Understanding in 2014, on expanding communication between ICANN and Chinese institutions (ICANN, 2014).

Even so, it remains difficult to quantify China's presence in ICANN. At the present stage, Chinese actors not only participate in ICANN initiatives but also rhetorically support its role. For instance, Nanni notes how despite a low profile during the stewardship IANA transition, China expressed public support to ICANN and multistakeholder model. An interesting case is provided at ICANN50 by Lu Wei, at that time Ministry for cyberspace. That event took place in London in 2014 and can be considered one of the starting points of the IANA stewardship transition (Nanni, 2021). After the IANA stewardship transition, the discussion became more articulated at least at the academic level. It is possible to find both scholars who suggest an alternative path of global Internet governance in which China create new areas and platform to develop its global internet governance vision (Li & Zeng, 2019) and other academics who suggest a more direct engagement of China into ICANN activities (Lin & Ren, 2017). From an institutional perspective, it is worth noting that, right after the transition, Guo Feng, China's governmental representative, was appointed vice chair of ICANN's GAC.

Reviewing China's participation at ICANN's activities, Lin and Ren two Tsinghua scholars who publish their commentaries also on CCP news portal diagnose a lack of consistent strategy, poor organisation and fragmented participation. Their study suggests that first, ICANN governance is based on bottom-up process, so its decision making process differs from UN organisations like ITU. China has not accepted this difference. Second, both before and after the break in ICANN GAC relations, China has always participated passively, and Chinese technical contribution has always been very limited. Third, the above mentioned withdrawal from GAC from 2001 to 2009 had a very negative effect in the global discussion on Internet governance (Lin & Ren, 2017).

The IANA transition presented an important step to develop a new Chinese approach on Internet governance. Indeed, the two Chinese scholars suggest moving away from a vision juxtaposing the multistakeholder versus multilateral model, as these two models are not necessarily in conflict. This requires the Chinese government to play a role in the background, leaving more space to technicians, engineers, scholars and think thanks. Furthermore, the transition opens space to replicate the same pattern of Chinese participation in the ITU that is increasing the engagement in terms proposals addressed by delegates and technicians. To enhance China's role in ICANN, they advocate a more active Chinese presence at ICANN activities as well as a more defined and leading role in the ICANN governance through the organisation of ICANN sponsored meetings. This process requires the creation of a competitive national network of technicians and engineers serving to increase Chinese influence on the global internet discussion through to the development of new theories, initiatives, and actions. They maintain an optimistic view on the overall role ICANN, which will become an increasingly inclusive and open organisation (*jiang chengwei yige yue lai yue kaifang baorong de guoji zuzhi*). Despite the two scholars' optimism, there are no concrete evidences and suggestions on how Chinese new theories, initiatives, and actions will be implemented.

## CHINA AND THE WORLD INTERNET CONFERENCE

China's approach to global Internet governance has not remained limited to joining existing processes and institutions, it is also increasingly taking initiatives to develop an original Chinese vision through the creation of its own platform and a direct engagement of foreign companies and high-level representatives of international organisations. A watershed moment happened in 2014 with the establishment of the Cyberspace Administration of China and the first edition of the World Internet Conference in Wuzhen. This annual event supports the Chinese vison to create an alternative framework on global Internet governance through the elaboration of new values such as "cyber sovereignty," through an inclusive approach addressed especially to the developing countries form the Global South, enriched by the dedicated project as the "Digital Silk Road," and through the engagement of international corporations and former ICANN delegates. Illustratively, the "Wuzhen

Declaration" presented at the first edition effectively was a draft joint statement supporting the idea that every nation has its own right to develop, use and govern the Internet as it sees fit. The document was criticised in circles such as the Internet Governance Forum Members Advisory Group (Aizu 2014) not only because of its content but also because it was slid under the doors of attendees' hotel doors the last night before the closing ceremony. At the second World Internet Conference, Xi Jinping presented once again, his vision of "cyber sovereignty" according to which the global Internet governance should "respect the right of individual countries to independently choose their own path of cyber development, model of cyber regulation, and internet public policies, and participate in international cyberspace governance on equal footing."

Furthermore, the World Internet conference was an important occasion to present to an international audience a new Chinese international governance theory based on Five Propositions (*wu dian zhuzhang*) (respect for cyber sovereignty, maintenance of peace and security, promotion of opens and cooperation, cultivation of good order) to create a cyberspace of shared destiny through the advancement of Four Principles (si xiang yuanze) (speed up the building of global Internet infrastructure and promote inter-connectivity; build an online platform for cultural exchange and mutual learning, promote innovative development of cyber economy for common prosperity, maintain cyber security and promote orderly development).

The role of the World Internet Conference was to challenge the *status quo* of the global Internet governance, and to propose a Chinese-led alternative to many countries. The organisers claimed that in 2015, "the World Internet Conference became truly global" through the engagement over of two thousand delegates form 120 countries and 20 international organisations (Thussu, 2018). Furthermore, compared to the "Wuzhen declaration," the "Wuzhen Initiative 2015," the official document published at this second World Internet Conference, supported a more inclusive and cosmopolitan approach to justify the Chinese vision on cyber sovereignty. Indeed, according to Shi, the second edition of the World Internet Conference expresses a vision that softens the nationalistic approach of China's advocacy on cyber sovereignty, whereas it highlights new keywords and expression such as the call to build "a community of common destiny, a concept that intertwines the classical Chinese philosophy of *Tianxia* (all under Heaven) with the Euro-American concept of cosmopolitanism" (Shi, 2017). If the "Wuzhen declaration" mainly reflects the need to defend cyber security and intellectual property, the "Wuzhen Initiative" at least from a narrative point of view, invests on the idea of co-governance, which can be considered a basic feature of the community of common destiny. In 2015 the Chinese government also issued a white paper introducing the idea of the "Information Silk Road," a strategy framed within the Belt and Road Initiative (BRI), aimed at creating synergies with BRI countries on emerging technologies for development and trade (Shen 2018; Bora 2020). The "Information Silk Road" was further elaborated into the concept of "Digital Silk Road" during the fourth World Internet Conference in 2017. A document published at this conference was co-signed by BRI partners like Laos, Egypt, Turkey, Thailand, Saudi Arabia, and Serbia outlining specific components focused on the improvement of broadband access, the promotion of digital technologies, the development of e-commerce capabilities and the promotion of international standards. Although some studies argue that the formalisation of these initiatives was below Beijing exceptions (Triolo et al. 2020), the World Internet Conference had a leading role in promoting Chinese global ambitions aimed at mitigating industrial overcapacity, facilitating corporate China's global expansion, constructing a China-cantered transnational network infrastructure, and promoting an Internet-enabled "inclusive globalisation" (Shen 2018). These goals have not been achieved yet also because of COVD-19, US-China trade war.

The WIC has also sought to attract high-profile participants. Previous ICANN CEO Fadi Chehadé became co-chair of the event's oversight committee in 2016. In 2017, Apple made its first appearance at the conference as its CEO Tim Cook gave a keynote speech. A Qualcomm senior officer hold a speech on the future of 5G standards and Artificial Intelligence. Even Bob Kahn, who is considered one of the fathers of the Transmission Control Protocol (TCP) and Internet Protocol (IP), delivered a speech at the event. The appointment of Mr. Chehadé, as well as the engagement with US private corporations, can be interpreted as an effort to improve the reputation of the event at the international level but also as an attempt to facilitate the discussion with Western agents in the field of global Internet governance. This effort is further illustrated by more recent appearance of ITU delegates at the World Internet Conference. In 2017 ICANN delegate Sally Costerton considered an opportunity "to interact with different stakeholder groups to raise awareness of ICANN and multistakeholder model" (Costerton 2017). In 2019, Malcolm Johnson, ITU Deputy Secretary, delivered a speech in which he clarified that ITU counted on China as a major partner, reiterating its gratitude to the Chinese government for its strong support to ITU (ITU 2019). At the present stage, there is not empirical evidence about concrete changes in the global Internet governance caused by the past editions of the World Internet Conferences. However, it still important to note how the international engagement has been growing at least until the fifth edition of the event. Indeed, in 2018 the conference registered the direct engagement of five international organisations such as the United Nations Department of Economic and Social Affairs, the ITU and the World Intellectual Property Organisation. Beside the political dimension, it also should be noted that that from the fifth edition, the World Internet Conference host collateral events addressed to the private sector. This is the case of "The light of the Internet Expo" an international event hosted in Wuzhen during the World Internet Conference, joined by more than eighty enterprises and aimed at facilitating the exchange between Internet companies. Also in this case, although these initiatives do not concretely impact the development of Internet governance neither in the narrow or broader sense, they are still useful to see how China has increased its confidence at the international level also in the field of the Internet and its willingness to actively influence its future trends.

It remains to be seen whether the World Internet Conference will maintain its global ambition and attractiveness at the international level. Even before the US-China trade and the COVID-19 pandemic, the conference experienced a decrease in terms of attendees, especially from America's biggest tech companies (Lahiri 2018). That said, the World Internet Conference can still be seen as a Chinese attempt to propose an alternative platform aimed at presenting its vision of Internet governance raising its discursive power in this specific domain (xianshi chu zhongguo zai hulianwang lingyu huayu quan de tigao) (Li and Zeng 2019).

## CONCLUSION

This chapter presented three different arenas in which China engages with global Internet governance discussions: ITU, ICANN, and the World Internet Conference. ITU is still the international organisation that most reflects China's preferred views, based on the concept of cyber-sovereignty and the role of the state. It this sense, it should not be surprising the fact that, at the time of writing, China's presence at ITU is relevant. Indeed, beside the presidency of Mr. Zhao as secretary general, two study group of the ITU-T are chaired by Chinese delegates.

However, ICANN is still one of the most important international organisations with a higher impact of the global Internet governance. This chapter shows how China changed its relations moving from a lack of official of communications, refuting to send its delegation to join GAC meetings to develop new forms of cooperation like the establishment of the first Engagement Centre in Beijing aimed at facilitating the collaboration Chinese authorities in 2013 but also expressions of public support to ICANN and the multistakeholder model like it happened during ICANN50 in London. This shift is important because it is now possible to argue that China has largely accepted the role of ICANN, suggesting its vision on global Internet governance is more complex than the simple notion of interstate multilateralism.

Finally, this chapter argues that to further promote its own vision, China created the World Internet Conference as a new platform for discussion that, in eight years, shifted its approach from the promotion of a defined vision on cyber sovereignty to the support of a more inclusive and less normative approach. Indeed, after the case of the "Wuzhen declaration," most of the topics discussed in the last editions of the annual World Internet Conference emphasise keywords such as "mutual trust" and "collective governance," more in line with the current multi stakeholder model and, at least apparently, in contrast with the "cyber sovereign" (Shi 2017). That said, we still need time and empirical evidence to understand to what extent this attitude will be concrete and sincere.

These three arenas and their relations with China remain uncertain especially after the US–China trade war and the COVID-19 pandemic. Coming to the ICANN case, this article shows how the debate on the China's role is polarised: if, on the one hand, two influential Chinese scholars on ICANN and close to the CCP's line suggest a more active role, others support the idea of an alternative platform as the World Internet Conference, which, however, witnessed a decrease of engagement from US companies in the last few years because of COVID-19 and US–China trade war. All in all, although China raised its voice, presence, and activities in the global discussion, it still has not changed the status quo of the global Internet governance. In the coming years it will be crucial to see further developments of the Chinese presence within ITU, its contributions in the field of new standards recommendations as well as its role in influencing different working and study groups. This new stage will not see the engagement of Mr. Zhao Houlin who ended its second mandated in September 2022. At the same time, it will be interesting to note how ICANN–China relationships will develop both in China and at the international level and whether China will maintain its positive attitude on multistakeholder model. Finally, the new editions of the World Internet Conferences will tell us whether its role will remain focused on a discursive domain or whether (and eventually how) it will gain a real and concrete power in the global Internet governance process.

## NOTES

1. In 2008 it became Ministry of Industry and Information Technology (MIIT).
2. One of the three supporting organisations called for in the ICANN Bylaws.

## REFERENCES


Aizu, I. (2014). *[IGFmaglist] WIC and My reservation to Wuzhen Declaration*. http://intgovforum.org/pipermail/igfmaglist_intgovforum.org/2014-November/002327.html

Arsène, S. (2015). "Internet domain names in China. articulating local control with global connectivity." *China Perspectives* 4, 25–34.

Bauer, J. M., and Dutton, W. H. (2015). *The New Cybersecurity Agenda*.

Bora, L. (2020). Challenge and perspective for digital Silk Road. *Cogent Business & Management* 7(1), 1804180.

Bygrave, L. A., and Bing, J. (2009). *Internet governance: Infrastructure and institutions*. OUP Oxford.

CAC (2016) 《国家网络空间安全战略》全文, 12月27日 http://www.xinhuanet.com//politics/2016-12/27/c_1120196479.htm.

Chen, Yin. (2009, November 18). Taking Stock and looking forward. *ICANN YouTube Channel*. Retrieved from https://www.youtube.com/watch?v=Ou1cAUXOluc.

Chenou, J.-M. (2014). From cyber-libertarianism to neoliberalism: Internet exceptionalism, multi-stakeholderism, and the institutionalisation of internet governance in the 1990s. *Globalizations* 11(2), 205–23.

CNBC (2016). *Domain name marketplace Sedo says 54% of new gTLDs are owned by the Chinese*. (2016). https://www.cnbc.com/2016/06/02/domain-name-marketplace-sedo-says-54-percent-of-new-gtlds-are-owned-by-the-chinese.html.

Costerton, S. (2017). *ICANN in Wuzhen, China—Fourth World Internet Conference and More*. https://www.icann.org/en/blogs/details/icann-in-wuzhen-china--fourth-world-internet-conference-and-more-29-11-2017-en.

DeNardis, L., & Raymond, M. (2013). Thinking clearly about multistakeholder internet governance. *GigaNet: Global Internet Governance Academic Network, Annual Symposium*.

DNSO. (2002, October 29). Worldwide alliance of top-level domain: Name communiqué from Shanghai meeting. Retrieved from http://www.dnso.org/constituency/cctld/docs/20021029.ccTLDshanghai-communique.html.

Drolet, E. (2022). Proposals at ITU-T for Internet Evolution Raise Serious Concerns, According to ISOC, *NANOG*, 4 August https://www.nanog.org/stories/new-ip-proposals-are-a-threat-according-to-isoc/.

Epstein, D. (2013). The making of institutions of information governance: The case of the Internet Governance Forum. *Journal of Information Technology* 28(2), 137–49.

FMPCR. (2015). *Remarks by H. E. Xi Jinping President of the People's Republic of China At the Opening Ceremony of the Second World Internet Conference*. https://www.fmprc.gov.cn/eng/wjdt_665385/zyjh_665391/201512/t20151224_678467.html.

Galloway, T. (2015). *China & technical global internet governance: From norm-taker to norm-maker?* Ph.D. Thesis. Deakin University.

Guan, T. (2019). The 'authoritarian determinism'and reductionisms in China-focused political communication studies. *Media, Culture & Society* 41(5), 738–50.

Harvey, D. (2005). A brief history of neoliberalism. Oxford: Oxford University Press.

Hu, Qiheng. (2004) Speech on the Consultation Meeting on the Establishment of the UN Working Group on Internet Governance," Geneva, 23–25 November.

Hughes, C. R., and Ermert, M. (2003). *What's in a name?: China and the domain name system*.

黄旭 (Huang, Xu). (2016). 我国参与全球互联网治理组织的过程和动力分析——以互联网工程任务组为例. 湖南科技大学学报: 社会科学版 19(5), 128–33.

ICANN & DoC. (1998). Memorandum of understanding between the US Department of Commerce and Internet corporation for assigned names and numbers. http://www.icann.org/en/general/icann-mou-25nov98.htm.

ICANN (2002). DNSO Structure https://archive.icann.org/en/dnso/dnso.htm

ICANN46 (2013) ICANN Engagement Center to Open in Beijing https://www.icann.org/resources/press-material/release-2013-04-08-en.

ICANN, GAC. (2013). *ICANN46 Beijing Communique*. https://gac.icann.org/content-Migrated/icann46-beijing-communique.

ICANN. (2014, June 23). LONDON—Welcome ceremony and president's opening. Retrieved from https://archive.icann.org/meetings/london2014/en/schedule/mon-welcome/transcript-welcome-23jun14-en.pdf.

IGF. (2012). *Main Sessions 2012 | Internet Governance Forum*. https://www.intgov-forum.org/multilingual/content/main-sessions-2012.

ITU (n/a). ITU of ITU-T Sector Members https://www.itu.int/online/mm/scripts/gensel11?_sect=T.

ITU. (s.d.). *Sixth Wuzhen World Internet Conference—Opening*. ITU. Retrieved on 27 June 2022 from https://www.itu.int:443/en/osg/dsg/speeches/Pages/2019-10-20.aspx.

ITU (2019). *Opening Speech by Malcom Johnson, ITU Deputy Secretary General*, 29 October https://www.itu.int/en/osg/dsg/speeches/Pages/2019-10-20.aspx.

Kleinwacher, W. (2004). Beyond ICANN Vs ITU? How WSIS tries to enter the new territory of Internet governance. *Gazette (Leiden, Netherlands)* 66(3–4), 233–51.

Klimburg, A. (2013). The Internet Yalta. Center for a New American Security. American Society, 2. Retrieved from https://www.cnas.org/publications/reports/the-internet-yalta.

Lahiri, T. (2018). *Silicon Valley is less visible at China's Wuzhen internet conference this year—Quartz*. https://qz.com/1455436/silicon-valley-is-less-visible-at-chinas-wuzhen-internet-conference-this-year.

Li Yan (李彦) & Zeng Runxi (曾润喜). (2019). 中国参与国际互联网治理制度建构的路径比较. 当代传播 5.

Lin Han (刘晗) & Ren Qiming (任启明). (2017). 简析如何在 ICANN 新机制下维护我国网络主权. 中国信息安全 5, 43–5.

Lindsay, J. R., Cheung, T. M., and Reveron, D. S. (2015). *China and cybersecurity: Espionage, strategy, and politics in the digital domain*. New York: Oxford University Press.

Mathiason, J. (2009). *Internet governance: The new frontier of global institutions*. London: Routledge.

MacKinnon, R. (s.d.). *China @ ICANN: Thoughts from former CEO Paul Twomey*. RConversation. Recuperato 27 giugno 2022, da https://rconversation.blogs.com/rconversation/2009/07/china-icann-thoughts-from-former-ceo-paul-twomey.html.

Mueller, M. L. (2011). China and global Internet governance: A tiger by the tail. *Access contested: Security, identity, and resistance in Asian cyberspace*, 177–94.

Nanni, R. (2021). The 'China'question in mobile Internet standard-making: Insights from expert interviews. *Telecommunications Policy* 45(6), 102151.

Negro, G. (2020). A history of Chinese global Internet governance and its relations with ITU and ICANN. *Chinese Journal of Communication* 13(1), 104–21.

Negro, G. (2022). China's Perspective on Internet Governance: a more Integrated Role in the Global Discussion? *Journal of Chinese Political Science*, 1–21.

O'Connor, J. (2014, August 23). The battle for soft power: America's digital cold war with China. Huffington Post. Retrieved from https://www.huffingtonpost.com/justinoaconnor/america-china-digital-war_b_5523110.html.

PRC State Council. (2015). *Initiative offers road map for peace, prosperity*. http://english.www.gov.cn/archive/publications/2015/03/30/content_281475080249035.htm.

Scribd. (s.d.). *World Internet Conference Draft Declaration | PDF*. Recuperato 27 giugno 2022, da https://it.scribd.com/document/247566581/World-Internet-Conference-Draft-Declaration.

Sha, Zukang (2003). "Statement by Head of the Chinese Delegation" WSIS PreConference 1, Geneva, 1–5.

Shen, Hong. (2016). China and global internet governance: Toward an alternative analytical framework. *Chinese Journal of Communication* 9(3), 304–24.

Shen, Hong. (2018). Building a digital silk road? Situating the internet in China's belt and road initiative. *International Journal of Communication* 12, 19.

Shi, Anbin. (2017). China's role in remapping global communication. In *China's media go global*. Routledge, 34–51.

Suttmeier, R. P., Yao, X., and Tan, A. Z. (2009). Standards of power? Technology, institutions, and politics in the development of China's national standards strategy. *Geopolitics, History, and International Relations* 1(1), 46–84.

Thussu, D. (2018). A new global communication order for a multipolar world. *Communication Research and Practice* 4(1), 52–66.

Triolo, P., Allison, K., Brown, C., and Broderick, K. (2020). The Digital Silk Road: Expanding China's Digital Footprint. *Eurasia Group* 8.

WSIS (World Summit on the Information Society). (2005). Tunis agenda for the information society, WSIS-05/TUNIS/DOC/6(Rev. 1)-E, 18 November 2005.

Xinhua. (2017). *Full Text: International Strategy of Cooperation on Cyberspace—Xinhua | English.news.cn*. 3 January, http://www.xinhuanet.com//english/china/2017-03/01/c_136094371_2.htm.

Yan, Li. (2015). Reforming Internet Governance and the Role of China. *Focus Asia* (12).

Zhang, Qiang. (2019). *China's Internet Governance: A New Conceptualization of the Cyber-Sovereignty Model*.

Zhao, Houlin. (2007). Internet governance: A personal perspective. *Information Polity* 12(1–2), 39–47.

Zhu, Hongbin. (2012). "The Impact of the New gTLD Program on the Internet Governance Regimes of Emerging Economies-China's Domain Name Regulation Revolution." In *GigaNet: Global Internet Governance Academic Network, Annual Symposium*.

# *Chapter 8*

# Becoming a Cyber Superpower

## *China Builds Offensive Capability with Military, Government, and Private Sector Forces*

Mei Danowski

Over the past thirty years, China's information and communications technology (ICT) sector has seen explosive growth. China has become the world's largest ICT exporter since 2004 (Ning Lutao 2009). The ICT sector is also the largest manufacturing sector within the Chinese economy, representing 55 percent China's GDP in 2021 (China—Technology and ICT 2022). As China's ICT sector grows, so has China's investment and focus on cybersecurity. The three-year (2021–2023) cybersecurity industry development plan, published in July 2021 by China's Ministry of Industry and Information Technology (MIIT), aimed to grow the industry to $39 billion by 2023, over 15 percent compound annual growth rate (Xiong Xinyi Zhang Hongpei 2021). China's pursuit of offensive cyber capability parallels the development of the country's ICT. Both developments are part of China's goal of becoming a "cyber superpower (*wangluo qiangguo*)," which China defines as being on a par with the United States in cyberspace (Kania, Sacks, Webster, and Triolo 2017).

There are many definitions of offensive cyber capabilities. This paper defines a nation state's offensive cyber as the capability of breaching an adversary's computer systems to carry out disruptive, destructive, or psychological effects in cyberspace to achieve strategic goals (Moor, 2022; Austin, Tay, and Sharma 2022; Smeets and Lin 2018; JP3-12 2018; Zetter 2022). This definition can include surveillance carried out to facilitate military action but, for the purpose of this paper, does not include purely cyber espionage that benefits China's economy more broadly, except those intelligence preparation for offensive activity. A nation state building its offensive cyber capabilities seeks to acquire "resources, skills, knowledge, operational concepts and procedures" that enable it to have an effect in cyberspace, as stated in a study from the Australia Strategic Policy Institute (Uren, Hogeveen, and Hanson 2018). China has been developing information warfare doctrines since the 1980s (Defense Refence 2016). However, it was the self-described "cyberwar" in 2001, waged by Chinese patriotic hackers against countries perceived to be doing harm to China (Smith C. 2001), that likely alerted Chinese leaders they needed to develop offensive capabilities they could control in line with their strategic priorities and superpower goals.

The Chinese military has been at the epicentre of China's offensive capability development, with the highest concentration of offensive cyber capabilities reside and important doctrinal development since the 1990s and concrete reforms since 2011. However, equally important are the efforts of other government agencies and of private individuals and companies who cooperate with the military. The patriotic hackers of 2001 eventually became private entrepreneurs and powered China's most innovative ICT developments (CCTV 2017). They have become valuable resources for building Chinese offensive capabilities as well, for example, through the cyber militia force building effort that began in the mid-2000s. These private companies develop valuable tools for military use through the nation's civil-military fusion strategy. Private cyber security companies do their part by discovering vulnerabilities and developing exploits, flaunting their capabilities at competitions such as the Tianfu Cup. Over time, China built up its offensive cyber capabilities by utilising a variety of human resources, from military and government personnel to civilian technology companies and other entrepreneurs.

Chinese military strategists have acknowledged the country's diverse cyber warfare forces, as evidenced by the 2013 version of the *Science of Military Strategy*, which outlined the basic structure of Chinese offensive cyber capabilities (Science of Military Strategy 2013). The book states that the popularity of the internet and the characteristics of military and civilian use determine the diversity of cyber offensive and defensive forces. China has three types of cyber warfare forces, according to the book (Military Strategy Institute, Military Battle in Cyber Domain 2013):

- Military combat forces specialising in cyber offensive and defensive operations. The book describes as professional military cyber warfare forces (*jundui zhuangye wangluozhan liliang*).
- Forces that specialised in cyber warfare and are formed within relevant government departments, such as the Ministry of Public Security (MPS) and the Ministry of State Security (MSS), and some other

PLA-authorised local forces. The book calls these "PLA-authorised forces (*shouquan liliang*)."[1]

- Individuals and entities, including private sector companies, carry out network attacks and defence when they are needed. The book refers to these civilian forces (*minjian liliang*) as "spontaneously" engaging in network attack and defence but can be organised and mobilised to conduct network operations. As we describe below, the private sector companies contribute to Chinese offensive capability either under contract with military and government or by developing capabilities that are brought to the military's attention and subsequent use.

The three types of cyber warfare forces display a clear picture of the different roles that military, government, industry, and hacker communities play in China's cyber warfare capabilities.

## BUILDING CYBER FORCES ONE STEP AT A TIME

Cyber warfare is a subset information warfare, according to Chinese military doctrine (Qian Fengshui 2004; ISCCC n.d.). Information warfare, including cyber, electronic, intelligence, and psychological warfare, has been a concern for the Chinese military since at least the late 1980s, before the internet was publicly available (Defence Refence 2016; Intelligence Warfare 2022). Chinese military strategists presented the concept of information warfare (IW) as a direct result of the dawning information era (Pan Ting 2005). However, it

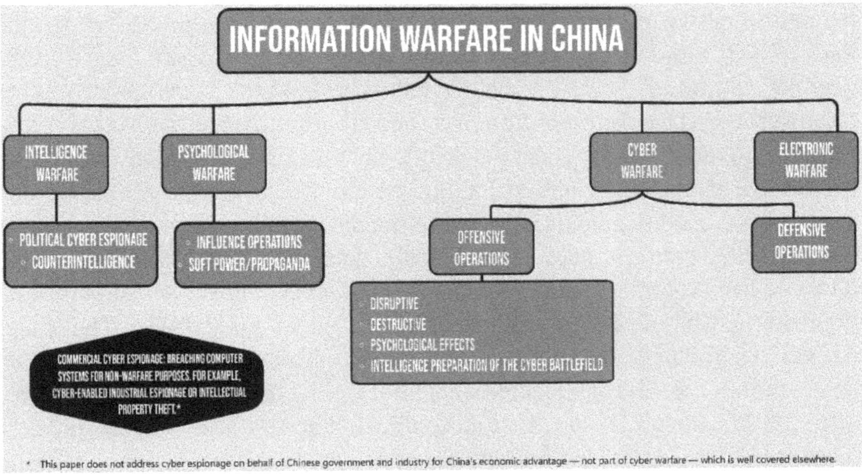


**Figure 8.1. In the Chinese context, China building its offensive capability is part of its cyber warfare capability under the umbrella of information warfare.**

was purely theoretical. Chinese Army Major Shen Weiguang published "The Rise of Information Warfare" in the *People's Liberation Army (PLA) Daily* on April 17, 1987, and published the book *Information Warfare* (*xinxi zhan*) in 1990. The book claimed the dawn of the information era would inevitably lead to information warfare. Unlike conventional conflicts, such a war would be waged on the battlefields of information network systems, using information weapons that can both destroy enemy information systems and influence the psychology of the adversary's population. The book calls for utilising information technology to complement military weaponry and equipment and for "occupying the high ground" in the battlefield of the information war (Li Qingshan 2002; Krekel, Adams, and Bakos 2012).

Events around the turn of the millennium convinced Chinese leadership of the need to develop information warfare capabilities they could control. In 1996, internet access became officially available to the public in China (Evolution of Internet in China 2001). Shortly after that, from 1999 to the early 2000s, during times of geopolitical tension, Chinese patriotic hackers waged what they termed a "cyberwar" against official websites in the United States, Japan, and Taiwan with disruptive denial-of-service (DoS) attacks or rudimentary website defacements (Aljazeera 2022). These state-encouraged patriotic hackers carried out their own form of offensive cyber operations to defend China against a perceived "attack." For example, in April 2001, a Chinese PLA Navy fighter jet pilot died in a mid-air collision with a US spy plane (CNN 2001). The infamous Honker Union hacker group led disruptive cyberattacks targeting hundreds of US websites, including those of the White House and California Department of Justice (Harris 2001). Interestingly, the Chinese government sometimes distanced itself from these hacktivists, claiming to disapprove of their operations. *The People's Daily*, China's Communist Party (CCP) newspaper, called these activities "web terrorism" and "unforgivable" (Smith, C. S. 2001). This is likely because the government wanted to have more state-controlled offensive capabilities. Since this was the first time the government denounced patriotic hackers openly, it likely wanted to portray the patriotic hackers as having gone out of control and to deny any government encouragement. The government remained silent on activities conducted by patriotic hackers previously, such as when they tried to "take down North Atlantic Treaty Organisation (NATO) networks" after NATO bombed the Chinese Embassy in Belgrade in 1999 (Wired 1999). The leader of the Honker Union claimed they had "achieved" their goal and called for an end (Smith, C. S. 2001). Hacktivism activities in China gradually died down after 2002. Though it reined in the patriotic hackers, the Chinese government seemed to realise the importance of offensive cyber capability, judging from subsequent government policy statements and actions. From then on, the Chinese government undertook a concerted effort to develop personnel structures and resources for offensive use.

## CYBER MILITIAS: A CYBER WARFARE RESERVE FORCE FOUNDED ON TECHNOLOGY COMPANIES

China started to experiment with using civilian resources, particularly the capabilities of technology companies, to build cyber militia forces as early as 2005. Some of these companies were founded by the same patriotic hackers who were involved in the self-claimed "cyberwar" only a few years earlier (Beech 2013). Building cyber militia forces was one of the efforts.

In 2005, the PLA organised a cyber militia unit at Chinese technology company Nanhao Group, located outside Beijing. This remained secret at first, only reaching the public in a 2011 *Financial Times* (FT) article (Hillie 2011). The cyber militia unit consisted of two groups tasked with offensive and defensive cyber operations. By participating in cyber militia forces, technology companies could "become part of the information warfare complex," in Hillie's words (2011). Some Chinese internet experts commenting on the 2011 FT report, speculated that companies like Nanhao Group did not conduct cyber operations with advanced techniques and likely carried out only "entry-level" network attacks such as HTTP flooding distributed

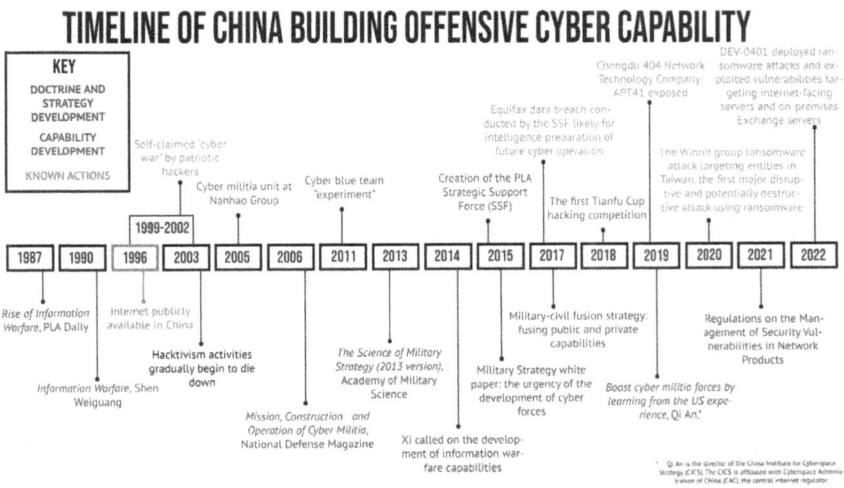


**Figure 8.2. The timeline of China building offensive cyber capability suggests doctrine and strategy development and force capability development have been parallel over past three decades while China putting its capability into practice happened more recently over last six years.**

denial-of-service (DDoS) attacks (Wu Yu 2011), the cyber militia unit at Nanhao group indicates the beginning of the cyber militia experiment.

The doctrinal justification for cyber militias appeared in public only after this secret experiment. In 2006, three officers from the Jiangsu Provincial PLA command's mobilisation department published a paper in *National Defence*, the magazine of the Academy of Military Science (AMS), suggesting the establishment of cyber militia units within ICT companies and scientific research institutions. The paper detailed the proposed cyber militias' missions, construction, and operations. Tasks included "stealing, changing, and erasing data" on enemy networks and intruding into those networks with the goal of "deception, jamming, disruption, throttling, and paralysis" (Li Guoqiang 2006).

As China's cyber security industry continued expanding, Chinese government statements again focused public attention on cyber militias. In November 2017, for example, *PLA Daily* reported that Harbin Garrison Commands, an armed police force, established a cyber militia unit at Antiy Technology company (Antiy), a leading cyber security company located in Harbin (Qin An 2019). The armed police forces are parts of the PLA, which are established in major cities in charge of military mobilisation and security (Armed Police Force 2021). In Antiy's case, its co-location of a cyber militia unit within a technology company was a clear example of military-civilian cooperation to build cyber capabilities.

In January 2019, Qin An, the director of the China Institute for Cyberspace Strategy (*Zhongguo wangluo kongjian zhanlue yanjiu suo*) discussed how China should learn from the US experience to build a cyber militia force (Qin An 2019). Qin referred to the US Navy's Navy Cyber Warfare Development Group (NCWDG) reserve unit, inaugurated 4 January 2019, whose stated mission is to draw on reservists' skills to help the NCWDG research and develop cyber, cryptologic, and electronic warfare capabilities (Naval Technology 2019). Qin pointed out that the NCWDG reservists were not formally part of the military but bolstered US cyber warfare capabilities. In addition, Qin cited US classified documents that former intelligence operative Edward Snowden had leaked, saying that most of the major US technology companies represented a military reserve force for the United States. Qin stressed that China's cyber militia efforts had lagged by comparison. China's cyber warfare reserve force should play a critical role in times of political, economic, and military "complexity," Qin wrote, referring both to confronting foreign adversaries and maintaining domestic order (Qin An 2019).

## INFORMATION WARFARE PLAN INTO ACTION

The world became aware of China's development of offensive cyber forces only in 2011 (Martin 2011), although as discussed previously, the country had begun developing private sector "cyber militias" since the early 2000s. The Chinese People's Liberation Army (PLA) publicly announced an "experiment" with building cyber forces in 2011.

In June 2011, the PLA built a "cyber blue team" that it said would help "safeguard internet security." It issued a statement saying, "internet safety has become an international issue" and noting that China suffers internet attacks from abroad (Martin 2011). As Western media questioned whether the cyber blue team could be an offensive force, *People's Daily* quoted the military experts' comments with the intent to clarify the official line. Major General Luo Yuan, deputy secretary-general of the Chinese AMS, said that the cyber blue team was just a code name for military training focused on cyber defence. Li Li, a military expert at China's Defence University, said that compared with the cyber forces of western countries, China's cyber blue team was in its infancy. It was not so much an organised and large-scale cyber warfare force but a military training model for "online confrontation" (wang shang duikang) (Guo, Gu, and Wu 2011). Traditionally, the cyber security industry defines a blue team as a team that plays a defensive role to defend against attacks while a red team plays an offensive role as attackers by finding vulnerabilities and breaking through cyber defence (Red Team/Blue Team Approach n.d.). Li's description about the online confrontation training model suggests a red team or an offensive team likely existed but the government did not publicly disclose its existence.

In the meantime, Chinese military strategists studied operational concepts and outlined procedures for cyber warfare including offensive cyber operations. In 2013, the AMS published a new version of *the Science of Military Strategy* (zhanlue xue) (Military Strategy Institute, The Science of Military Strategy 2013). It was the first time that a Chinese military publication addressed cyberwarfare holistically (Lyu 2019). The book recognised cyber offensive and defensive operations as the most important form of military battle in the cyber field. It stated that the main purpose of offensive and defensive network operations is to destroy the enemy's network systems and network information, while protecting one's own network system and network information. Whether conducting defensive or offensive cyber operations, the book posited, practitioners need deep familiarity with the working principles of the network they are either defending or attacking. They need to be able to access a specific network system, discover the defects and vulnerabilities in the system, and either exploit the vulnerabilities quickly, in an offensive setting, or patch them, in a defensive setting (Military Strategy Institute, The Science of Military Strategy 2013).

A high point in the development of information warfare came in September 2014, when Chinese President Xi Jinping, who also headed the Central Military Commission, officially called on the military to develop "a new military doctrine, institutions, equipment systems, strategies, tactics and management models" for information warfare. Xi called for promoting military innovation, changing "fixed mindsets" of traditional warfare, and establishing "the ideological concept of information warfare" (China Daily 2014). Following Xi's call for an information warfare plan, in May 2015, China's Military Strategy white paper emphasised the urgency of the development of cyber forces, "as cyberspace weighs more in military security" (State Council Information Office 2015). This military strategy white paper heralded the coming of PLA Strategic Support Force, a new combat force.

## Strategic Support Force: An Organised Cyber Force

Xi Jinping's vision for national defence and military reforms, including the development of cyber forces, resulted in the creation of the PLA Strategic Support Force (SSF) on December 31, 2015. The SSF consolidated the PLA's space warfare (*taikong*), signals intelligence, network offense and defensive cyber operations (*wangluo gongfang*), and electronic warfare (*dianzi duikong*) capabilities (Qiu Yue 2016). The core of SSF was described as a "new quality combat capability" (*xinzhi zuozhan nengli*) and "information system-based system of systems operational capability" (*jiyu xixin xitong de tixi zuozhan nengli*) (People's Daily 2015).[2] Chinese official military commentators have portrayed the SSF as an "information umbrella" (*xinxi san*) for the military system that provides "accurate, efficient, and reliable information as well as strategic support" (Qiu Yue 2016). The SSF regrouped operational units from the former General Armament Department (GAD) together with the network systems, electronic warfare, and technical reconnaissance department of the former General Staff Department (GSD). The SSF's establishment indicates the Chinese government's resolution to have an "informatised" (*xinxi hua*) and "world-class" military that can prevail in modern information warfare (Lin Kongshi 2016). The SSF's primary missions and functions—to provide information support, information warfare, and force development—and its organisation and personnel structure suggest that the SSF plays a leading and perhaps coordinating role in the PLA's cyber operations.

After its establishment in the tail end of 2015 the SSF conducted the Equifax data breach from May to July 2017. Although this cyber intrusion did not have the nature of a destructive or disruptive attack, the large volume of the data the PLA hackers obtained likely provided intelligence for future cyber operations. The US Department of Justice (DOJ) charged four members of the Chinese PLA in this case, which was only the second time that a DOJ indicted Chinese military personnel for hacking since 2014 (Department of Justice 2014).

On February 10, 2020, the DOJ unsealed an indictment charging four members of the Chinese PLA with hacking into the computer system of the credit reporting agency Equifax from May to July 2017 (DOJ 2020). The indictment states that the four PLA hackers were members of the PLA's 54th Research Institute. These four PLA hackers obtained 145 million items of personally identifiable information (PII) of Americans, 10 million Americans' driver's license numbers, 200,000 credit card numbers, and other PII. In addition, the PLA hackers obtained close to 1 million pieces of PII belonging to United Kingdom and Canadian citizens (DOJ 2020).

The indictment stated that the four PLA hackers were "member of the PLA's 54th Research Institute, a component of the Chinese military." Judging the PLA's organisation structure, the "PLA's 54th Research Institute" likely refers to the SSF's 54th Research Institute, formerly the PLA GSD Forth Department's 54th Research Institute (Stokes, Lin, and Hsiao 2011).

Further research indicates that the PLA's SSF 54th Research Institute also operates in civilian guise as the Beijing-based Northern Research Institute of Electronic Equipment of China (*Zhongguo beifang dianzi shebei yanjiu suo*) (NRIEEC). A biography of a deputy director of the 54th Research Institute, appearing in a Chinese-language website from the Harbin Institute of Technology Alumna Association includes one of his job titles as "Deputy Director of Northern Institute of Electronic Equipment of China (GSD 54th Research Institute)" (Harbin Institute of Technology Alumni Assoc 2020). Chinese-language internet searches yield little information about the NRIEEC. Searches in Chinese-language business information repositories have not produced any business registration information either. It is a common practice for Chinese military research institutes to have equivalent civilian sounding institute names to disguise military affiliations (Sharma 2018).

## CIVIL MILITARY FUSION STRATEGY: BUILDING OFFENSIVE CYBER CAPABILITIES IS NOT JUST A MILITARY EFFORT

China leapt into the internet era in the early 2000s when many Chinese technology companies sprang up. As described above, although the Chinese government initially denounced the actions of early patriotic hackers as "overly enthusiastic," these same hackers later became part of the establishment after their started companies as entrepreneurs (Tencent Security Labs 2017).

China's leaders encouraged them to become part of the establishment, recognising the importance of the dual use of technologies and attracting civilian forces to build cyber capabilities with military benefits.

When Xi Jinping called for an information warfare plan in 2014, he urged the integration of military and civilian innovation so the two sectors could accommodate each other and develop together (China Daily 2014). After the military reform gradually rolled out beginning in 2016, China elevated its military-civil fusion strategy to a new level with the establishment of the Central Commission for Integrated Military and Civilian Development (*Zhongyang junmin ronghe fazhan weiyuanhui*) in 2017, led by Xi Jinping.

On June 20, 2017, Xi Jinping spoke at the first plenary meeting of the new commission. According to the official Xinhua news agency, Xi said China's civil-military fusion policy should "value socialism's advantages of pooling resources to solve major problems and improve work efficiency." Xi said, "The ideas, decisions, and plans of military and civilian integration must be fully implemented in all fields of national economic and defence construction," including cyberspace (Xinhua News 2017).

Under the civil-military fusion strategy, many companies, particularly cyber security companies, were recruited or actively participated in various civil-military fusion projects. Companies such as Antiy Technology Group Co., Ltd. (*antian keji*), were named as part of the national team of cyber security. A PLA unit recognised Antiy for providing technical support and network security services during a satellite launch (Antiy 2021). In addition, leading cyber security companies such as Qihoo 360 Technology Co. Ltd (China National Radio 2017), Beijing Zhidaochuangyu Information Technology Co. Ltd (Knwonsec n.d.), NSFOCUS Technologies Group Co. Ltd (Allia Z-Park Joint Innovation Civil-Military Integration Equipment Industry Alliance 2021), Qi An Xin Technology Group Inc (Qianxin Innovation Teams n.d.), and Topsec Technologies Group Inc (Topsec Tech Group 2018) have military-civil fusion centres or participate projects related to China's military-civil fusion strategy.

## From "All People Are Soldiers" (*quanmin jie bing*) to "Extremely Lean" (*ji qi jinggan*): Cyber Combat Forces Require Highly Skilled Personnel

As noted above, the PLA reorganisation resulted in the creation of an organised and large-scale cyber warfare force, the Strategic Support Force (SSF). The military-private sector fusion strategy provided a channel for ICT companies to participate in projects, which enhance the military's cyber warfare capabilities. The development of the cyber militia force likely turned some ICT companies into part of the cyber warfare complex. As an integral part of this complex, highly skilled ICT practitioners have been indispensable in building China's cyber combat forces because they are extremely capable forces, as in the example of Chengdu 404, below.

The authors of the 2013 *Science of Military Strategy* noted that cyber warfare has a "broad mass base (*qunzhong jichu*)," likely referring to the earlier patriotic hacker activities. However, they stated it is impossible to achieve the traditional Communist Chinese military strategy of "all people are soldiers (*quan min jie bing*)" in cyberspace. This is because network offensive and defensive operations require specialised practitioners who are extremely capable. The category of "civilian forces" cited in the book likely refers to the talents from cyber security companies which play an important role in the military-civil fusion strategy and the development of cyber militia forces.

China's rapid technological development pushed many cyber security companies to recruit the best talent and promote innovation. Specialists from these companies are part of those "extremely lean" groups of capable practitioners that the *Science of Military Strategy* cites. These cyber security companies are among those civilian forces that the government and the military often mobilise to conduct network operations. As described more fully below, Chengdu Silingsi (404) Network Technology Company is one of these examples.

## Chengdu 404 Network Technology Company: Advanced Persistent Threat (APT) 41

On September 16, 2020, the US Department of Justice (DOJ) released a report detailing three separate indictments (DOJ 2020). Two of these indictments, one occurring August 2019 and the other in August 2020, charge five Chinese individuals with the computer intrusions of more than one hundred companies located in the United States and abroad. The indictments attributed the intrusions to APT41 (aka BARIUM), a cyber threat group security companies have tracked under various names (Fraser, Plan, and O'Leary 2019). Three of five individuals the indictment named—Jiang Lizhi, Qian Chuan, and Fu Qiang—were leaders of Chengdu 404, a network security company based in Chengdu, Sichuan province. Within Chengdu 404, Qian Chuan was president, Jiang Lizhi served as vice president for the Technical Department, and Fu Qiang served as manager for Big Data Development.

Examination of the company's website and business registration information shows that Chengdu 404's business resembled the role of a red team or an offensive team. Established in May 2014, Chengdu 404 claimed its services included penetration testing, APT attack monitoring, firmware trojan detection, mobile device forensics, research and products related to password recovery and anonymous proxy. The business partners listed in Chengdu

404's website included state-owned enterprises, universities, and government agencies related to information security (UMISEN n.d.).

Chengdu 404 appears to be one of the top cyber security companies in Sichuan Province. The Chengdu Information Network Security Association, a local industry association, named the company one of the outstanding companies in 2019. In December 2019, the Sichuan Bureau of the National Administration of State Secrets Protection awarded the Class B qualification of software development for confidential information system to Chengdu 404 which allowed the company to engage in classified state projects.

Chengdu 404 also demonstrated its capabilities by developing proprietary software and patents. According to Chengdu 404's business registration information, the company owns four patents and fifteen software copyrights. The most recent patent, a platform for processing dark net intelligence was registered on July 10, 2020 (QCC 2022).

Three indicted hackers from Chengdu 404 had appeared in local media as technologists with visions and patriotic spirit. In October 2018, *Sichuan Economic Daily*, a provincial government newspaper, published an interview with key personnel of Chengdu 404 (*Sichuan Economic Daily* 2018). The interview explained these hackers were "not typical hackers," but "hidden Chengdu white hats who take things seriously." The hackers claimed they were not crass "businessmen," but gentlemanly "entrepreneurs." and they alluded to the classical Chinese saying about "certain things that a gentleman would do, or not do (*junzi you suowei yousuo bu wei*)." They appeared to hold themselves to a high standard, aspiring to contribute to society and national security while also making their own technological dreams come true.

After the DOJ's disclosure of APT41's indictments, Chengdu 404 did not stop its operations. The company's hiring posts continued appearing at various Chinese recruitment platforms (BOSS 2022).

## "Vulnerability Should Be Considered as National Strategic Resource"

Vulnerability refers to "a weakness in an information system, system security procedures, internal controls, or implementation that could be exploited or triggered by a threat source" (Computer Security Resource Center n.d.). Since the 2013 version of *the Science of Military Strategy* described the importance of discovering and quickly exploiting vulnerabilities in information systems, China's technology elites across the cyber security industry expressed concerns over China's own vulnerability to cyberattacks and described ways the country could prepare for cyberwarfare. Zhou Hongyi, the CEO of Qihoo 360, the largest cyber security company in China, proposed the country should treat vulnerabilities as national strategic resources.

In September 2017, at the 5th China Internet Security Conference, Zhou Hongyi stated that cyber warfare is unavoidable, and vulnerabilities are weapons of cyber warfare. Zhou explained that the essence of cyber warfare is vulnerability. "This concept goes beyond technical vulnerabilities in software and encompasses broader weaknesses in whole networks. Understanding these network vulnerabilities, in both broad and narrow senses, is essential for building a network of weapons." Indeed, said Zhou, "vulnerabilities should be considered as national strategic resources" (Sina Photo News 2017; The Paper 2017).

Zhou Hongyi's speech about vulnerabilities as weapons of cyber warfare appears to be a turning point for China in recognising vulnerability discovery and exploitation as central to their offensive cyber capability and to the country's overall pursuit as a leading cyber power.

## THE TIANFU CUP: SHOWCASING OFFENSIVE CYBER CAPABILITY AND DEPTH OF OFFENSIVE CYBER INVENTORIES

Shortly after the September 2017 Internet Security Conference, China prohibited Chinese security researchers from participating in international hacking competitions in early 2018. This move made it easier for the Chinese government to control and retain vulnerability information inside the country (Bing 2018). In November 2018, Chinese technology giants including Alibaba, Tencent, and Baidu founded China's own international hacking competition, the Tianfu Cup (N 2018). In the 2018 Tianfu Cup, a team from 360Security, a subsidiary of Qihoo360, won first place. The team discovered and successfully exploited zero-day vulnerabilities from Apple Safari, iPhone X, Google Chrome, Microsoft Edge, Microsoft Office, and Oracle Virtual Box, since then, the Tianfu Cup competition target list has focused on foreign products (N 2018). At the 2021's Tianfu Cup, teams continued to focus on popular Western products such as Windows 10, Microsoft Exchange Server, Chrome, VMware workstation, and iPhone 13 Pro. In the meantime, the Tianfu Cup has drawn more attention from the Chinese government. In 2021 a cyber security summit, held as part of the hacking competition, attracted participants in the security field from military, central and local governments, government research institutes, and the Ministry of Public Security (Xinhuanet 2021). The choice of foreign products for the competition list of the Tianfu Cup encourages the discovery of vulnerabilities that Chinese strategists or military cyber forces can exploit.

The Chinese government values vulnerabilities so highly that it requires Chinese researchers do not divulge the vulnerabilities they discover until after they have informed Chinese government authorities. As a result, Chinese military or security personnel have an opportunity to exploit these vulnerabilities for use against domestic or foreign adversaries before defenders can patch them (Smalley 2022). In July 2021, the Ministry of Industry and Information Technology (MIIT), the Cyberspace Administration of China (CAC), and the Ministry of Public Security (MPS) published the "Regulation on the Management of Security Vulnerabilities in Network Products" (*wangluo changpin Anquan loudong guili guiding*), which came into effect on September 1, 2021. This regulation outlines how network product security vulnerabilities are discovered, reported, patched, and disclosed. It obligates network product suppliers to report the vulnerability to MIIT Network Security Threat Information Sharing Platform within two days of discovering a vulnerability in their product. In turn, the MIIT Network Security Threat Information Sharing Platform simultaneously reports vulnerabilities to the vulnerability platforms of the National Network and Information Security Alerting Centre. Article 9 of the Regulation prohibits providing information on undisclosed network product vulnerabilities to foreign organisations or foreign individuals, except for the network product supplier (CAC 2021).

The first high-profile case enforcing this regulation occurred in December 2021, involving Alibaba Cloud, one of the top cloud providers in China. China's MIIT suspended a cybersecurity partnership with Alibaba Cloud for six months after it failed to report Log4j vulnerabilities to MIIT first and instead reported it to the software provider Apache Software Foundation (Greig 2021). MIIT's action indicated the government's commitment to tightly controlling vulnerabilities and served as a warning to other technology companies to follow the rules or suffer the consequences.

Chinese researchers announced vulnerabilities in VMware products at the October 2021 Tianfu Cup, but VMware did not release patches for these vulnerabilities until February 2022, almost four months later. VMware's patch announcement indicated "these vulnerabilities were reported to the Chinese government by the researchers that discovered them, in accordance with their laws" (VMWare Blog 2022). This suggested the Chinese researchers followed the regulation on the management of security vulnerabilities by delaying their reporting to VMware until well after they reported to the Chinese government.

At the same time, Chinese nation state cyber threat actors have taken advantage of vulnerabilities for strategic use. *MIT Technology Review* reported in May 2021 that an Apple device vulnerability discovered at the 2018 Tianfu Cup had been used in Chinese cyber espionage campaigns against the Uyghurs, the Chinese Muslim minority, two months before the vulnerability was reported and fixed (O'Neill 2021). A July 2020 DOJ indictment alleged two Chinese Ministry of State Security (MSS) affiliated threat actors, Dong Jiazhi and Li Xiaoyu, targeted US biotechnology, pharmaceutical, and medical companies, seeking COVID-19 related research and trade secrets. The indictment revealed the threat actors received a zero-day exploit from an email sent by an MSS officer (DOJ 2020). This same tactic can be used in any offensive cyber operation.

The Tianfu Cup competition demonstrates China's offensive cyber capabilities to hold key Western systems and networks at risk and highlights the substantial depth of China's offensive cyber inventories (Work 2021). Military and government agencies alike can use these capabilities for disruptive, destructive, or psychological operations against foreign or domestic targets.

## PUTTING OFFENSIVE CYBER CAPABILITY INTO PRACTICE

China claims its national cyber security strategy is to maintain active defence, defined as a combination of strategic defence and actively preparing for offensive attacks (Xinhua News, Active Defense Strategy 2015). This differs from the approach of countries such as the United States and many of its allies, which explicitly prescribes going beyond defence to develop offensive cyber forces and cyber deterrence strategy (Lu Chuanying 2019). At the 2021 World Internet Conference, also known as the Wuzhen Summit, a global conference organised by the Cyberspace Administration of China, Chinese President Xi Jinping presented China's solution to "build a strong digital security barrier" (shuzi anquan pinzhang) to ensure cyber security (He Yin 2021). However, China's Defence White Paper in 2019 also advocated the building of cyberspace capabilities that are "consistent with China's international standing as a major cyber power," thereby implying the necessity of building offensive capabilities as well (Ding Yang 2019). China has used offensive cyber resources in cyber espionage and, increasingly, in other destructive operations.

The US government published alerts with lists of vulnerabilities used by Chinese state-sponsored threat actors, often in particular combinations for greater potency. One such alert from October 2020 listed twenty-four publicly known vulnerabilities that Chinese state-sponsored threat actors had exploited against various network and communication systems and devices (National Security Agency Cybersecurity Advisory 2020). In June 2022, the US government warned that Chinese actors were using well-known, but inconsistently patched vulnerabilities to breach firewalls and other elements of communications networks to gain a foothold throughout essential communications infrastructure (Cybersecurity Advisory 2022).

Many security firms also illustrate how Chinese state-sponsored actors exploited vulnerabilities. In March 2021, Microsoft detected a threat campaign conducted by HAFNIUM, a group assessed to be state-sponsored operating out of China, using multiple zero-day exploits to attack on-premises versions of Microsoft Exchange Server (Microsoft Threat Intelligence Center 2021). In March 2022, Mandiant reported that APT41, a Chinese state-sponsored group, used zero-day vulnerabilities in the USAHerds application and in the Log4J logging application to target US state government networks (Brown, Ta, and Bienstock 2022).

Chinese state-sponsored groups have in some striking recent incidents used destructive and disruptive tools. In May 2020, Taiwan Ministry of Justice Investigation Bureau (Investigation Bureau) reported targeted attacks on several Taiwan-based petrochemical companies and one semiconductor manufacturing plant halted operations and forced the companies to isolate the affected networks and restore backup files (Investigation Bureau 2020). The Investigation Bureau attributed the ransomware attack to a China-based group called the "Winnti group" (Staff writer 2020). Security company Trend Micro analysed the ransomware family and indicated the attack was potentially destructive rather than merely disruptive, as "the ransomware appeared to target databases and email servers for encryption" (Trend Micro 2020). If it was not intended to be reversed in return for the payment of ransom, this points to a political rather than a financial motivation and implies the perpetrators were state sponsored.

This was the first major destructive attack using ransomware by a Chinese state-sponsored group in recent years. Chinese cyberthreat actors often use Taiwan as a testing ground because of the common language. In addition, the Chinese perception that Taiwan is rightfully part of China that world powers will not retaliate against China for aggression against a diplomatically isolated Taiwan.

In January 2022, Microsoft reported another China-based ransomware operator, DEV-0401, deployed multiple ransomware attacks and exploited vulnerabilities in internet-facing systems running Confluence and on-premises Exchange servers. In one campaign, DEV-0401 exploited a vulnerability targeting internet-facing servers running vulnerable instances of VMware Horizon. After successful intrusions, the actor deployed the NightSky ransomware (Microsoft Defender Threat Intelligence 2021). Researchers from SecureWorks, tracking DEV-0401 as BRONZE STARLIGHT, assessed the short lifespan of each ransomware family the actor deployed in the attacks suggested the actor was using ransomware as a smokescreen to cover its cyber espionage or intellectual property theft activities (Paganini 2022).

Nation state actors sometimes use ransomware attacks for political reasons, to disrupt or destruct target organisations, or to clean up or cover the traces of cyberespionage. In China's case, testing the capabilities of ransomware attacks is relatively new, but will likely continue.

## WHAT'S NEXT?

Since Chinese President Xi Jinping publicly promoted the importance of cybersecurity for the country in 2014 with the slogan "without cybersecurity there will be no national security; without informatisation, there is no modernisation" (Wang Yang 2014). Xi has repeatedly given high-level talks and speeches with directives on how to ensure "cybersecurity as the important part of national security." Most of Xi's talks emphasise building cyber power for defensive purposes and he prefers phrases such as "the protection of information infrastructure" and "the construction of cybersecurity incident response command capabilities" (Creemers, Triolo, and Webster 2018). Xi Jinping has not openly discussed China's desire to build offensive cyber capabilities. However, the statement in the 2019 Defence White Paper that the country's cyber capabilities must be equivalent with China's international standing as a major cyber power implies the country also needs an offensive cyber capability.

In conclusion, China has been building its offensive cyber capability by integrating resources from the military, government, and ICT industries while making organisational changes, implementing regulations, and initiating national strategies to support the effort. Chinese military strategists have been studying and developing operational concepts and procedures related to information warfare for decades. To a certain degree, this has accelerated the process of building capable cyber combat forces. Rapid technological development in China has created a robust cyber security industry. The Chinese government considers talents from the cyber security industry are the most capable civilian forces to build its offensive cyber capability. These talents are in the forefront of the cyber field to help the nation stockpile vulnerabilities, one of the most effective cyber weapons, as well as develop exploits and place vulnerabilities in use.

The Winnti group that deployed the ransomware attack against organisations in Taiwan is likely just the beginning of China's cyber combat forces' use of destructive and disruptive tools. As the divide between China and the democratic world expands, becoming a cyber superpower with offensive cyber capability is essential for China to compete as a major power in the world.

## NOTES

1. Little public information is available on the details of the connection between the military and the MPS/MSS. These agencies likely mainly carry out espionage. The available evidence does not point to their taking an autonomous/major role in offensive cyber activity defined for this paper. Other than the 2013 Science of Military Strategy doctrine, there is little public information. This likely indicates an ongoing power struggle within the Chinese system between the PLA's leadership and the government agencies to determine who truly oversees Chinese action in cyberspace, as Joe McReynolds, a research analyst at Defence Group Inc., points out in a paper published at the Jamestown Foundation's China Brief (volume XV, no. 8) on April 17, 2015.

2. Systems of systems operational capability is the integration of C4ISR (command, control, communications, computers, intelligence, surveillance, and reconnaissance) and forces to significantly multiply war-fighting capacity and enable joint operation capability.

## REFERENCES

*Armed Police Force*. (2021, November 27). Retrieved from TengXun Wang: https://new.qq.com/omn/20211127/20211127A04QOY00.html.

Aljazeera. (2022, April 7). *Hacked: Inside the US-China Cyberwar*. Retrieved from Aljazeera: https://www.aljazeera.com/program/101-east/2022/4/7/hacked-inside-the-us-china-cyber-war.

Allia Z-Park Joint Innovation Civil-Military Integration Equipment Industry Alliance. (2021, December 17). *Lianmeng yu Lvmeng Keji Qianshu Zhanlue Hezuo Xieyi* (The Alliance and NSFOCUS Signed a Strategic Cooperation Agreement). Retrieved from Z-Park Joint Innovation Civil-Military Integration Equipment Industry Alliance Enterprise Service Platform: https://web.archive.org/web/20220608032032/http://39.105.31.242/union/lmnewsdetail?id=44.

Antiy. (2021, September 30). *Youli Baozhang Gaofen Wuhao Weixing Fashe Renwu Yuanman Chenggong Antian Shoudao Ganxiexin* (Antiy received a letter of thanks for the successful launch of the Gaofen-5 02 satellite mission). Retrieved from ANTIY: https://archive.ph/s3a7M.

Austin, G., Tay, K., and Sharma, M. (2022, February 24). *Great-Power Offensive Cyber Campaigns: Experiments in Strategy*. Retrieved from The International Institute for Strategic Studies (IISS): https://www.iiss.org/blogs/research-paper/2022/02/great-power-offensive-cyber-campaigns.

Beech, H. (2013, February 21). *China's Red Hackers: The Tale of One Patriotic Cyberwarrior*. Retrieved from Time: https://world.time.com/2013/02/21/chinas-red-hackers-the-tale-of-one-patriotic-cyberwarrior.

Bing, C. (2018, March 8). *China's government is keeping its security researchers from attending conferences*. Retrieved from CyberScoop: https://www.cyberscoop.com/pwn2own-chinese-researchers-360-technologies-trend-micro.

BOSS. (2022, March 31). *Chengdu 404 Hiring*. Retrieved from BOSS: https://archive.ph/dkuFu.

Brown, R., Ta, V., and Bienstock, D. (2022, March 8). *Does This Look Infected? A Summary of APT41 Targeting U.S. State Governments*. Retrieved from Mandiant: https://www.mandiant.com/resources/apt41-us-state-governments.

CAC. (2021, July 13). *Gongye he Xinxihua bu Guojia Hulianwang Xinxi Bangongshi Gonganbu Guanyu Yinfa Wangluo Chanping Anquan Loudong Guanli Guiding de Tongzhi* (MIIT, SIIO, and MPS Notice of Network Products Security Vulnerability Management Regulations). Retrieved from Cyberspace Administration of China (CAC): https://archive.ph/OywzR.

CCTV. (2017, July 30). *Zhongguo Shoubu Heike Jilupai "Wo shi Heike"* (China's First Hacker Documentary "I am a Hacker"). Retrieved from QQ Video: https://v.qq.com/x/page/t0531ltq698.html.

*China—Technology and ICT*. (2022, August). Retrieved from Privacy Shield: https://www.privacyshield.gov/article?id=China-Technology-and-ICT#:~:text=and%20trade%20data.-,Overview,(IT)%20consulting%20firm%20IDC.

China Daily. (2014, September 1). *Army needs 'information warfare' plan, declares Xi*. Retrieved from China Daily: https://archive.ph/0g4K0.

China National Radio. (2017, December 26). *Wangluo Kongjian Anquan Junmin Ronghe Chuangxin Zhongxin Zhengshi Chengli* (Cyberspace Security Civil-Military Fusion Innovation Center was Officially Established). Retrieved from CNR: https://web.archive.org/web/20220606171951/http://china.cnr.cn/gdgg/20171226/t20171226_524077343.shtml.

CNN. (2001, April 1). *US aircraft collides with Chinese fighter forced to land*. Retrieved from CNN.com: https://web.archive.org/web/20081211063330/http://archives.cnn.com/2001/US/04/01/us.china.plane.03.

Computer security Resource Center. (n.d.). *Vulnerability*. Retrieved from NIST: https://csrc.nist.gov/glossary/term/vulnerability.

Creemers, R., Triolo, P., and Webster, G. (2018, April 30). *Translation: Xi Jinping's April 20 Speech at the National Cybersecurity and Informatization Work Conference*. Retrieved from New America Cybersecurity Initiative: https://www.newamerica.org/cybersecurity-initiative/digichina/blog/translation-xi-jinpings-april-20-speech-national-cybersecurity-and-informatization-work-conference.

Cybersecurity Advisory. (2022, June 7). *NSA, CISA, and FBI Expose PRC State-Sponsored Exploitation of Network Providers, Devices*. Retrieved from National Security Agency/Central Security Service: https://www.nsa.gov/Press-Room/News-Highlights/Article/Article/3055748/nsa-cisa-and-fbi-expose-prc-state-sponsored-exploitation-of-network-providers-d.

Defence Refence. (2016, 1 6). *"Guang zhanzheng": 30 nian hou de Zhanzheng xin Xingtai* (the "War of Light": A new form of warfare 30 years later). Retrieved from Xinhua Net: https://archive.ph/1GMXO.

Department of Justice (DOJ). (2014, May 19). *U.S. Charges Five Chinese Military Hackers for Cyber Espionage Against U.S. Corporations and a Labor Organization for Commercial Advantage*. Retrieved from United States Department of Justice:

https://www.justice.gov/usao-wdpa/pr/us-charges-five-chinese-military-hackers-cyber-espionage-against-us-corporations-and.

Department of Justice (DOJ). (2020, February 10). *Chinese Military Personnel Charged with Computer Fraud, Economic Espionage and Wire Fraud for Hacking into Credit Reporting Agency Equifax*. Retrieved from United States Department of Justice: https://www.justice.gov/opa/pr/chinese-military-personnel-charged-computer-fraud-economic-espionage-and-wire-fraud-hacking.

Department of Justice (DOJ). (2020, September 16). *Seven International Cyber Defendants, Including "Apt41" Actors, Charged in Connection with Computer Intrusion Campaigns Against More Than 100 Victims Globally*. Retrieved from The United States Department of Justice: https://www.justice.gov/opa/pr/seven-international-cyber-defendants-including-apt41-actors-charged-connection-computer.

Department of Justice (DOJ). (2020, July 21). *Two Chinese Hackers Working with the Ministry of State Security Charged with Global Computer Intrusion Campaign Targeting Intellectual Property and Confidential Business Information, Including COVID-19 Research*. Retrieved from the US Department of Justice: https://www.justice.gov/opa/pr/two-chinese-hackers-working-ministry-state-security-charged-global-computer-intrusion.

Ding Yang. (2019, July 24). *Xin Shidai de Zhongguo Guofang Baipishi* Quanwen (Full text of the White Paper on China's National Defence in the New Era). Retrieved from Ministry of National Defence of the People's Repubic of China: https://archive.ph/ZaCtJ.

*Evolution of Internet in China*. (2001, January 1). Retrieved from China Education and Research Network: https://web.archive.org/web/20120527120608/http://www.edu.cn/introduction_1378/20060323/t20060323_4285.shtml.

Fraser, N., Plan, F., and O'Leary, J. (2019, August 7). *APT41: A Dual Espionage and Cyber Crime Operation*. Retrieved from MANDIANT: https://www.mandiant.com/resources/apt41-dual-espionage-and-cyber-crime-operation.

Greig, J. (2021, December 22). *Chinese regulators suspend Alibaba Cloud over failure to report Log4j vulnerability*. Retrieved from ZDNET: https://www.zdnet.com/article/log4j-chinese-regulators-suspend-alibaba-partnership-over-failure-to-report-vulnerability.

Guo, L., Gu, C., and Wu, N. (2011, June 27). *Zhuanjia tan Zhongguo Zujian Wangluo Landui Yuanyin Zhizai Baozhang Wangluo Anquan* (Experts talk about the reasons for China to form a cyber blue team aimed at ensuring cyber security). Retrieved from www.chinanews.com: https://archive.ph/pFc2c.

Harbin Institute of Technology Alumnni Association. (2020, September 29). *Lv Yueguang*. Retrieved from Harbin Institute of Technology Alumnni Association: https://archive.ph/HxcmX.

Harris, S. (2001, May 2). *Chinese hackers declare war on US web sites*. Retrieved from Government Executive: https://www.govexec.com/technology/2001/05/chinese-hackers-declare-war-on-us-web-sites/9062.

He Yin. (2021, September 29). *Bawo Jiyu Xieshou Goujian Wangluo Kongjian Mingyun Gongtongti* (Seize the Opportunity to Build a Community of Destiny in Cyberspace Together). Retrieved from qstheory.cn: https://archive.ph/6kLKX.

Hillie, K. (2011, October 12). *Chinese military mobilises cybermilitias*. Retrieved from Financial Times: https://www.ft.com/content/33dc83e4-c800-11e0-9501-00144feabdc0.

*Intelligence Warfare*. (2022, January 20). Retrieved from Encyclopedia of China: https://www.zgbk.com/ecph/words?SiteID=1&Name=情报战&Type=bkzyb&subSourceType=000003000011000001.

Investigation Bureau. (2020, May 15). *Guonei Zhongyao Qiye Zao Lesou Ruanti Gongji Shijian Diaocha Shuoming* (Investigation of Ransomware Attack on Important Domestic Enterprises). Retrieved from Taiwan Ministry of Justice Investigation Bureau: https://archive.ph/Vc5qw.

ISCCC. (n.d.). *Jiedu Guoneiwai Wangluozhan Xingshi* (Interpreting the Situation of Domestic and Foreign Cyber Warfare). Retrieved from China Cybersecurity Review Technology and Certification Center: https://www.isccc.gov.cn/xwdt/xwkx/04/253384.shtml (https://archive.ph/bbmYI#selection-509.0-509.10).

JP3–12. (2018, June 8). *Joint Publication JP 3–12, Cyberspace Operations.* USCYBERCOM. Retrieved from https://irp.fas.org/doddir/dod/jp3_12.pdf.

Kania, E., Sacks, S., Webster, G., and Triolo, P. (2017, September 25). *China's Strategic Thinking on Building Power in Cyberspace*. Retrieved from DigiChina Stanford University: https://digichina.stanford.edu/work/chinas-strategic-thinking-on-building-power-in-cyberspace.

Knwonsec. (n.d.). *Milestones*. Retrieved from Knownsec.com: https://web.archive.org/web/20220617022132/https://www.knownsec.com/#/milestones.

Krekel, B., Adams, P., and Bakos, G. (2012). *Occupying the Information High Groud: Chinese capabilities for computer network operations and cyber espionage.* Washington DC: Northrop Grumman.

Li Guoqiang, C. W. (2006). *Minbing Wangluozhan Fendui de Renwu Jianshe yu Yunyong* (Mission, Construction and Operation of Cyber Militia Force). *National Defence Magazine*, 8.

Li Qingshan. (2002). Study of high technology war. In A. O. Sciences, *Military Science: Chinese Academic Canon in the 20th Century* (168). Fuzhou: Fujuan Education Publisher.

Lin Kongshi, H. J. (2016, August 30). *Xi Jinping Shicha de Zhanlue Zhiyuan Budui shi Yizhi Zenyang de Liliang* (What Kind of Force Is the Strategic Support Force that Xi Jinping Visited?). Retrieved from CCTV.COM: https://archive.ph/jmUKm.

Lu Chuanying. (2019, October 20). *Goujian Wangluo Kongjian Mingyun Gongtongti Youzhu yu Jiaqiang Guoji Wangluo Anquan Zhili* (Building a Community of Destiny in Cyberspace Helps Strengthen International Cybersecurity Governance). Retrieved from Sohu.com: https://archive.ph/fAe3z.

Lyu, J. (2019, April 1). *What Are China's Cyber Capabilities and Intentions?* Retrieved from Carnegie Endowment for International Peace: https://carnegieendowment.org/2019/04/01/what-are-china-s-cyber-capabilities-and-intentions-pub-78734.

Martin, R. (2011, May 27). *China Insists Cyber Blue Team is Temporary, for Defence*. Retrieved from TechAsia: https://www.techinasia.com/china-cyber-blue-team.

Microsoft Defender Threat Intelligence. (2021, December 11). *Guidance for preventing, detecting, and hunting for exploitation of the Log4j 2 vulnerability*. Retrieved from Microsoft Security: https://www.microsoft.com/security/blog/2021/12/11/guidance-for-preventing-detecting-and-hunting-for-cve-2021-44228-log4j-2-exploitation/#NightSky.

Microsoft Threat Intelligence Center. (2021, March 2). *HAFNIUM targeting Exchange Servers with 0-day exploits*. Retrieved from Microsoft Security: https://www.microsoft.com/security/blog/2021/03/02/hafnium-targeting-exchange-servers.

Military Strategy Institute. (2013). Military Battle in Cyber Domain. In M. S. AMS, *The Science of Military Strategy* (196). Beijing: Academy of Military Sciences.

Military Strategy Institute. (2013). *The Science of Military Strategy*. Beijing: Academy of Military Sciences.

Moor, D. (2022). *Offensive Cyber Operations: Understanding Intangible Warfare*. London: Hurst & Company.

N, B. (2018, November 20). *Tianfu Cup 2018 PWN—Ethical Hackers Hacked Apple, Adobe, Google, Microsoft, Oracle, VMware & Earned 1,000,000 USD*. Retrieved from GBHackers on Security: https://gbhackers.com/tianfu-cup-2018-pwn.

National Security Agency Cybersecurity Advisory. (2020, October). *Chinese State-Sponsored Actors Exploit Publicly Known Vulnerabilities*. Retrieved from media.defence.gov: https://media.defence.gov/2020/Oct/20/2002519884/-1/-1/0/CSA_CHINESE_EXPLOIT_VULNERABILITIES_UOO179811.PDF.

Naval Technology. (2019, January 9). *New Navy Cyber Warfare Development Group Reserve Unit Opens*. Retrieved from Naval Technology: www.naval-technology.com/news/cyber-warfare-development-reserve.

Ning Lutao. (2009). *China's Rise in the World ICT Industry: Industrial Strategies and the Catch-Up Development Model*. New York: Routledge.

O'Neill, P. H. (2021, May 6). *How China turned a prize-winning iPhone hack against the Uyghurs*. Retrieved from MIT Technology Review: https://www.technologyreview.com/2021/05/06/1024621/china-apple-spy-uyghur-hacker-tianfu.

Paganini, P. (2022, June 26). *China-linked APT Bronze Starlight is deploying post-intrusion ransomware families as a diversionary action to its cyber espionage operations*. Retrieved from Security Affairs: https://securityaffairs.co/wordpress/132624/apt/bronze-starlight-deploy-ransomware.html.

Pan Ting. (2005, January 1). *Duihua "Xinxi Zhan"* ("Information Warfare" Dialogue). Retrieved from Chinese Youth Daily: http://zqb.cyol.com/content/2005-01/22/content_1021198.htm.

People's Daily. (2015, Decemeber 20). *Tixi Zuozhan* (System Operations). Retrieved from Xinhuanet: http://news.xinhuanet.com/politics/2015-12/20/c_128548116.htm.

QCC. (2022, July 29). *Chengdu 404*. Retrieved from Qcc.com: https://archive.ph/cIQk4.

Qian Fengshui. (2004, July 22). *Junshi Pinglun: Jiedu Xinxizhan, Wangluozhan, Wangluo Zhongxinzhan* (Military Commentary: Interpreting Information Warfare, Cyber Warfare, and Network Centric Warfare). Retrieved from Sina Military: http://jczs.news.sina.com.cn/2004-07-22/2153212084.html.

*Qianxin Innovation Teams*. (n.d.). Retrieved from Qianxin: https://web.archive.org/web/20220620000031/https://www.qianxin.com/threat/threatsafeinstitute?tid=56.

Qin An. (2019, January 22). *Zhongguo Ying Ruhe Jiejian Meiguo Jingyan Dazao 'Wangluo Minbing'* (Boost Cyber Militia Forces by Learning from the US Experience). Retrieved from The Paper: http://m.thepaper.cn/yidian_promDetail.jsp?contid=2882486&from=yidian.

Qiu Yue. (2016, 01 05). *Zhuanjia Jiemi Zhanlue Zhiyuan Budui Shi Wojun Zai Taikong deng Zhanchang Qude Jubu Youshi* (Experts unveil strategic support forces to enable our military to achieve local advantage in space and other battlefields). Retrieved from The Paper: https://www.thepaper.cn/newsDetail_forward_1417087.

*Red Team/Blue Team Approach*. (n.d.). Retrieved from NIST Computer Security Resource Center: https://csrc.nist.gov/glossary/term/red_team_blue_team_approach.

Sharma, Y. (2018, October 29). *Scholar hide military links from Western universities*. Retrieved from University World News: https://www.universityworldnews.com/post-mobile.php?story=20181029193127483.

*Sichuan Economic Daily*. (2018, October 28). *Doyinyushi de "Baimao Heike" Fang Chengdu shi Silingsi Wangluo Keji Youxian Gongsi* (The Hidden "White Hat Hacker" Chengdu404 Network Technology Company). Retrieved from Read01: https://archive.ph/qB8df.

Sina Photo News. (2017, September 12). *Diwujie Hulianwang Anquan Dahui Zhaokai Zhou Hongyi Ren shi Da Anquan Shidai Hexin* (The 5th Internet Security Conference, Zhou Hongyi: People is the Core of Great Security). Retrieved from photo.sina.com.cn: https://archive.ph/PwKpN.

Smalley, S. (2022, August 10). *China could be reviewing security bugs before tech companies issue patches, DHS official says*. Retrieved from CyberScoop: https://www.cyberscoop.com/dhs-official-chinese-rules-exploit.

Smeets, M., and Lin, H. S. (2018). Offensive Cyber Capabilities: to What Ends. In T. Minarik, R. Jakschis, and L. Lindstrom, *10th International Conference on Cyber Conflict CyCon: Maximising Effects* (55–72). Tallinn: NATO CCD COE Publications.

Smith, C. (2001, May 13). *May 6–12; The First World Hacker War*. Retrieved from the *New York Times*: https://www.nytimes.com/2001/05/13/weekinreview/may-6-12-the-first-world-hacker-war.html.

Staff writer. (2020, May 17). *Bureau names ransomware culprits*. Retrieved from *Taipei Times*: https://www.taipeitimes.com/News/taiwan/archives/2020/05/17/2003736564.

Stokes, M. A., Lin, J., and Hsiao, L. R. (2011, November 11). *The Chinese People's Liberation Army Signals Intelligence and Cyber Reconnaissance Infrastructure.*

Retrieved from Project 2049 Institute: https://project2049.net/wp-content/uploads/2018/05/pla_third_department_sigint_cyber_stokes_lin_hsiao.pdf.

Strategic Research Department of AMS. (2013). Strategic Guidance for Military Struggles in Cyber Domain. In A. O. Science, *The Science of Military Strategy* (195). Beijing: Military Academic Works Academy of Military Science.

Tencent Security Labs. (2017, July 30). *Yangshi Xieshou Tengxue Anquan "Wo shi Heike" Jilupian Zhanxian Zhongguo Baimao Heike Fengcai* (CCTV and Tencent Security: "I am a Hacker" Documenentary to Show China's White Hat Hacker Style). Retrieved from ZhiHu: https://zhuanlan.zhihu.com/p/28197077.

The Paper. (2017, September 18). *360 Zhou Hongyi Wangluo Da Anquan Shidai Daolai Loudong Chengwei Guojia ji Zhanlue Ziyuan(360 Zhou Hongyi: Great Security Era Vulnerabilities Become National Strategic Resources)*. Retrieved from Sohu.com: https://archive.ph/ZVA3D.

The State Council Information Office. (2015, May). *China's Military Strategy.* Retrieved from James Town Foudation: https://jamestown.org/wp-content/uploads/2016/07/China%E2%80%99s-Military-Strategy-2015.pdf.

Topsec Tech Group. (2018, September 19). *Tianrongxin CEO Li Xueying Boshi Chuxi Wangluo Anquan Juemin Ronghe Fazhan Luntan* (Dr. Li Xueying, CEO of Topsec Attended the Forum on Civil-Military Fusion in Cybersecurity). Retrieved from Sohu: https://archive.ph/Dpeeg.

Trend Micro. (2020, May 6). *Targeted Ransomware Attack Hits Taiwan Organizations*. Retrieved from Trend Micro: https://blog.trendmicro.com/trendlabs-security-intelligence/targeted-ransomware-attack-hits-taiwanese-organizations.

UMISEN. (n.d.). *About Us*. Retrieved from UMISEN Chengdu 404: https://archive.ph/4XFbS.

Uren, T., Hogeveen, B., and Hanson, F. (2018, July 4). *Defining Offensive Cyber Capabilities.* Retrieved from Australian Strategic Policy Institute: https://www.aspi.org.au/report/defining-offensive-cyber-capabilities.

US Federal Bureau of Investigation. (2020). *Indicators of Compromise Associated with Cyber Intrusions and Malicious Acts Attributed to the People's Liberation Army (PLA), 54th Research Institute (RI).* FBI Flash AC-000121-TT.

VMWare Blog. (2022, February 15). *VMSA-2022–0004: Questions & Answers*. Retrieved from VMWare: https://core.vmware.com/vmsa-2022-0004-questions-answers-faq#sec19469-sub1.

Wang Yang. (2014, February 27). *Xi Jinping Zhuchi Zhaokai Zhongyang Wangluo Anquan he Xinxihua Lingdao Xiaozu Diyici Huiyi* (Xi Jinping Hosted the First Meeting of the Central Leading Group on Cybersecurity and Informatization). Retrieved from www.GOV.cn: https://archive.ph/sSBbx.

Wired. (1999, September 2). *China Fought Bombs with Spam*. Retrieved from Wired: https://www.wired.com/1999/09/china-fought-bombs-with-spam.

Work, J. D. (2021, October 22). *China Flaunts Its Offensive Cyber Power*. Retrieved from War on the Rocks: https://warontherocks.com/2021/10/china-flaunts-its-offensive-cyber-power.

Wu Yu. (2011, October 14). *Waimei Jie Zhongguo Junfang Guyong "Wangluo Minbing"* (Foreign Media Reveal that the Chinese Military Employs "Cyber

Militia"). Retrieved from DW.Com: https://www.dw.com/zh/外媒揭中国军方雇佣网络民兵/a-15460503.

Xinhua News. (2015, January 26). *Active Defence Strategy*. Retrieved from People.cn: https://archive.ph/Q9OVD.

Xinhua News. (2017, June 21). *Xi Jinping: Jiakuai Junmin Ronghe* (Xi Jinping: Speed up the Civil-Military Fusion). Retrieved from Beijign Youth Daily: http://epaper.ynet.com/html/2017-06/21/content_253346.htm?div=-1.

Xinhuanet. (2021, October 28). *"Tianfu Bei" 2021 International Cybersecurity Competition and Summit Forum Held in Sichuan)*. Retrieved from news.cn: https://archive.ph/8Lhpv.

Xiong Xinyi Zhang Hongpei. (2021, July 12). *China launches 3-year draft plan for cybersecurity sector after regulatory actions*. Retrieved from Global Times: https://www.globaltimes.cn/page/202107/1228461.shtml.

Zetter, K. (2022, June 17). *What It Means that the U.S. Is Conducting Offensive Cyber Operations Against Russia.* Retrieved from Zero Day: https://zetter.substack.com/p/what-it-means-that-the-us-is-conducting.

# PART IV

# Local Dynamics

# *Chapter 9*

# China—A Rising Tech Power?

## *National Ambitions and Local Realities*

Genia Kostka

In the battle for global tech dominance, China is rapidly surpassing its Western competitors (Olsen, 2020).[1] The country is quickly reaching global leadership in many areas of science and technology, including facial recognition, certain fields of AI and e-mobility. In this chapter, I argue that the rise is both fueled and constrained by the specific institutions of the party-state. The 'fuel' is the party-state's capacity and will to lead China up the value chain thanks to massive investments. The 'constraints' have to do with the downside of decentralised and fragmented authoritarianism.

This chapter begins by analysing China's growing technological power by juxtaposing national ambitions with local realities. Despite Beijing's impressive efforts to devise industrial policies for technology upgrading (Naughton 2021), there is a substantial high-tech policy implementation gap. The term 'implementation gap' refers here to differences between Beijing's high-tech ambitions and local policy outcomes. The reason for the gap may be that many elements of Beijing's tech agenda fall to local governments for delivery. As local governments' pre-existing industrial structures, interests, and capabilities differ widely, national plans and investment funds are often not (or only partially) implemented, poorly executed, or significantly delayed.

The analysis further shows that China's national technology policies and plans have been implemented unevenly across regions. By focusing on three provinces (i.e., Sichuan, Anhui, and Zhejiang), this analysis highlights how different institutional structures have shaped the provinces' technological development trajectories. A historical comparison sheds light on the diverse state–business relations in the high-tech industry: While Sichuan's tech industry has, to a large extent, been dictated by government and defence projects, tech companies in Zhejiang have benefitted from more freedom to develop. In Anhui, the tech industry has formed particularly close relationships with local research institutes and labs located in Hefei. The historical institutional perspective offered here helps to explain why the high-tech industry in Sichuan is largely focused on civil–military industry production, while in Anhui and Zhejiang there is a stronger focus on AI technology and speech recognition.

## NATIONAL AMBITIONS

China has seen astonishing technological advances in the past few decades. It has the largest 5G network and the most extensive optical fibre cable network in the world, and it is producing self-driving cars. It is already leading in many AI technologies, including AI-based emotion recognition and facial recognition technologies (Kharpal 2019). Two of the world's largest supercomputers—Tianhe-2 and Sunway TaihuLight—are also located in the country (Abbany 2017). Rapid advances have also been made in high-speed quantum computing. In 2016, China successfully launched its quantum satellite, Micius (or QUESS), the first in the world (Disha 2021). Researchers at the University of Science and Technology of China in Hefei recently announced a new quantum computing breakthrough that allegedly surpassed Google's achievements, making it the world's leader in quantum technology (Corbett and Singer 2022). On the Global Innovative Index, China climbed from twenty-ninth place in 2015 to twelfth in 2021 (World Intellectual Property Organisation 2022). These are impressive achievements, and there is no question that China is becoming a leader in global science and technology innovation.

However, China's industrial technology capabilities should not be overstated, and for many digital technologies, China is still catching up. Particular vulnerabilities are in the integrated circuit and basic software industries. In 1999, then Minister of Science and Technology Xu Guanhua famously said, 'The Chinese ICT industry lacks a core (chips) and souls (basic software)' (*Zhongguo xinxi chanye 'que xin shao hun*') (Bu 2020). Since then, China has invested massive sums in the semiconductor and software industries to increase domestic capacity, but it still relies largely on imports for high-end chips, which state media often describe as being 'wedged by the neck' (*Ka bozi*).

No matter how one assesses China's technological capabilities, there is general agreement that technological progress has been at the core of the political agenda for a very long time. The period following the global financial crisis in 2009 was especially significant as policymakers shifted from an indicative planning approach to new industrial policies in which the state plans to invest unprecedented amounts of money to leapfrog other nations in technology (Naughton 2021). The 2015 release of both *Made in China 2025* and *Internet Plus Strategies* marked a new stage in the state's efforts to lead the tech industry up the value chain. Other relevant policy initiatives that support China's tech rise include the *Action Outline for Promoting the Development of Big Data* (2015), *the Outline of National Informatisation Development Strategy* (2016) and *the Development Plan on the New Generation of Artificial Intelligence (AI)* (2017). In the *Plan on the Next Generation of AI*, the Chinese government outlines its road map to become the primary AI 'innovation centre' by 2030 (Webster et al. 2017). Between 2014 and 2020 alone, China's Industrial Guidance Funds (IGFs) raised a staggering US$1.6 trillion for the targeted sectors (Naughton 2021: 106),[2] thereby underlining the Chinese party-state's commitment to leaping ahead in strategically important technologies.

China's recent five-year plans (FYPs) also reflect the growing emphasis on tech primacy. The 13th FYP (2016–2020) called for the expansion of strategic emerging industries (SEIs) and opens its chapter on the National Big Data Strategy with a statement that the government 'will make big data a fundamental strategic resource . . . to help transform and upgrade industries and bring about innovation in social governance' (NDRC 2016). The 14th FYP (2021–2025) dedicated an entire section (section 5) exclusively to facilitating digitalisation and establishing a digital China, and it picked seven key emerging technologies to be further promoted to speed up the country's ambitious tech advancement, as well as ten sectors where the technologies are encouraged to be applied (NDRC 2021).[3]

Statements by national leaders further underline the political will of China's party-state to win the global race for technological leadership. At the Fourth Plenary Session of the 19th Central Committee of the Communist Party of China in 2019, leading cadres listed data as one of the seven major factors of production, along with labour, capital, land, knowledge, technology, and management. Xi Jinping also stressed China's national tech ambitions in the thirty-fourth collective study of the Political Bureau of the 19th Central Committee in 2021 when he stated: 'In today's era, digital technology and digital economy are the opportunities for the world's technological revolution and industrial transformation, and they are the key areas of a new round of international competition. We must seize the opportunities and seize the commanding heights of future development' (Xi 2022).

In addition to having comprehensive top-down planning and using massive funds to support homegrown tech companies, China has also developed a large-scale domestic talent promotion programme. In May 2020, the Ministry of Education launched the 'School of Future Technology' programme to upgrade technological and innovation capabilities through educational investments (Li 2021). A stated programme goal was to move Chinese innovation power from 'Made in China' (*Zhongguo zhizao*) to 'Created in China' (*Zhongguo chuangzao*) (Huang 2016). In the first batch, twelve universities were selected with strengths in critical technologies such as aerospace, AI, quantum information science, marine technology, and life and health science. The plan is to expand this to another twenty to thirty 'Schools of Future Technology' in the future. Table 1 lists the twelve higher-education institutions selected to implement the programme (Zhongguo Jiaoyu Zaixian 2021).

**Table 9.1: Twelve Universities Selected for the School of Future Technology Programme**

| | *Higher-Education Institution* | *Technologies* |
|---|---|---|
| 1 | Peking University, Beijing | Big Data and Biomedical Artificial Intelligence Department: biomedical imaging, molecular medical sciences, biomedical engineering, big data, and biomedical artificial intelligence |
| 2 | Tsinghua University, Beijing | Advanced chips, new materials, software, AI, intelligent manufacturing, and national security |
| 3 | Beihang University (BUAA), Beijing | Aerospace/aviation |
| 4 | Tianjin University (TJU), Tianjin | Smart/intelligent machines and systems, storage science and engineering, smart city, etc. |
| 5 | Northeastern University (NEU), Shenyang, Liaoning, in cooperation with Huawei | Control science and engineering, computer science and technology, software engineering, robotics |
| 6 | Harbin Institute of Technology (HIT), Harbin | AI, intelligent manufacturing, life and health sciences |
| 7 | Shanghai Jiao Tong University (SJTU), Shanghai | Energy and environment, health and medicine |
| 8 | Southeast University (SEU), Nanjing | Chip design, information materials, future communication, intelligent perception and sensing (*zhineng ganzhi* 智能感知) |
| 9 | University of Science and Technology of China (USTC), Hefei, Anhui | Quantum technology |
| 10 | Huazhong University of Science and Technology (HUST), Wuhan | Advanced intelligent manufacturing, biomedical imaging, photoelectron chips and system, AI |
| 11 | South China University of Technology, Guangzhou | intelligent perception and sensing, big data, AI+ technologies |
| 12 | Xi'an Jiaotong University (XJTU), Xi'an | AI, energy storage sciences and engineering, intelligent manufacturing, biomedical engineering, smart city |

Sources: Zou 2021; Li 2021; Zhongguo Jiaoyu Zaixian (eol.cn) 2021

Despite all the planning and investment, turning these tech ambitions into reality is a challenge. It is well known that in China's highly decentralised authoritarian structures (Landry 2008), local governments play a key role in shaping implementation outcomes, which often results in the 'selective' implementation of national policy (Li and O'Brien 1999). Often it is the local governments that have to create an attractive investment environment for innovation and research. For instance, many local governments have created special development zones and high-tech industrial parks, but not all of them were successful in creating the necessary conditions for high-tech industrial cluster growth (Kania and Laskai 2021). Many examples are known where local governments simply picked the wrong tech companies as a 'local champion' or where they overinvested in certain industries (Segal 2018), resulting in a duplication of efforts. In other words, Beijing strongly depends on provincial governments to support its tech agenda with the right means and tools.

The next section highlights how technological trajectories vary across regions. At the provincial level, the trajectory of tech advancement is often shaped by multiple pre-existing economic, social, and political factors. By looking at local governments' technical, financial, and political capacities to push for tech leadership in their locally grown tech industries, the final section will explain why national tech ambitions are often only partially implemented at the local level.

## HIGH-TECH SECTOR DEVELOPMENT AND REGIONAL PATH DEPENDENCY

China's national technology policies and plans have been implemented unevenly across regions. This is partly the result of local governments adjusting and repurposing national policies to make them fit the local context. Additionally, local trajectories for innovation and technological advancement depend heavily on pre-existing infrastructures and conditions. Therefore, the growth of regional tech hubs is very path-dependent in that new outcomes are firmly tied to previous outcomes rather than the current conditions alone (Isaksen 2015). Below, the focus will be on three provinces (i.e., Sichuan, Anhui, and Zhejiang) to illustrate how different existing institutional structures shaped their provincial trajectories in high-tech sector development.

### Sichuan

Located in Western China, Sichuan province is home to many car manufacturing plants and major high-tech suppliers of critical car manufacturing components, such as lithium batteries for Tesla and integrated circuit assembly for foreign companies like Intel, Texas Instruments, and Onsemi. Sichuan is also home to a large military and defence sector, with many research institutions and factories for military-use aircraft, rockets, and components for nuclear weapons headquartered there. Some of the high-tech products produced in Sichuan are dual-use technologies, that is, they partly or fully originate in the defence industry, which has made use of AI and other advanced technologies (Chen 2022).

Sichuan showcases how a national security movement in the 1960s set the stage for a strong linkage between a local defence industrial base and a growing civilian, high-tech economy. The origin of Sichuan's high-tech industrial sector is often linked to China's Western Development Strategy (*Xibu dakaifa*) in 2001. However, this explanation gives insufficient credit to industrial policies that can be traced back further—specifically, to the Mao era. In the early 1960s, during the Cold War, Mao proposed a geo-military industrial grand plan called the Third Front Movement (*Sanxian jianshe*), which started in 1964 and targeted mountain regions in southwestern and western parts of China for key military production. The isolated mountain areas were chosen as they would be the hardest for foreign forces to invade.

As a result of Mao's plan, large-scale investments were made in national defence complexes in the remote and mountainous areas of Sichuan. The new provincial military and defence sector included defence-related technology research, the transport sector, and other basic supporting industries such as manufacturing, mining, metal, and electricity supply. Table 2 provides an overview of key sectors developed during the Third Front Movement in Sichuan.

Although many Third Front plants went bankrupt after the 1980s because of bad planning, hasty implementation, and the geographical inaccessibility of supplies and markets, these areas retained a certain level of industrial infrastructure into the era of reform and opening. In the early 1980s, many companies and factories moved out of the mountainous areas to gain better access to the market and reinvented themselves to produce civilian goods rather than military supplies (Butterfield 1980). The purpose of this 'defence conversion' was to 'pull the military into the process of national macro-economic adjustment' (Lee 2011: 3). By 1996, almost all former military sectors, including the aviation and electronic industries, which formed industry clusters in Sichuan, were producing more than 80 percent of their total output on civilian products (Lee 2011: 4). Thus, the early industrial structures built during the Third Front provided fertile ground to grow a local electronic manufacturing sector in Sichuan province. In subsequent years, many of Sichuan's military firms diversified into the manufacturing sectors and even established joint ventures with foreign firms.

**Table 9.2: Industrial Sectors Developed During the Third Front Movement in Sichuan**

| *Sector* | *Location* |
|---|---|
| Industrial manufacturing | Chongqing*, Chengdu |
| Arms industry, including research institutions on defence, midsize to large enterprises specialising in the defence industry and civil–military enterprises | Chongqing* (production of conventional weapons such as rifles, tanks, trucks, and conventional powered submarines), Chengdu, Mianyang, Guangyuan, Leshan, Xichang, Daxian (now Dazhou) |
| Coal mining | Dukou (now Panzhihua; Panzhihua Iron and Steel), Guang'an Huaying (Lushuidong coal mine) |
| Petrochemical industry | Nanchong |
| Metallurgical industry | Dukou (Panzhihua), Daxian (Dazhou), Leshan, E'mei, Zigong, Jiangyou |
| Hydropower stations | Chengdu, Deyang, Gongzui, Zigong, Chongqing* |
| Machinery and electronic plants | Chengdu, Deyang, Mianyang, Jiangyou, Guangyuan, Leshan, Xichang, Zigong, Neijiang, Luzhou, Ya'an, Fuling Dist. (Chongqing)*, Wanxian (Chongqing)*, Guang'an Huaying |
| Aviation and aerospace industry | Chengdu and satellite cities such as Deyang/ Guang'han (Civil Aviation Flight University of China), Ya'an, Mianyang, Xichang and Daxian (Dazhou) |
| Nuclear industry | Mianyang (research, the Chinese Academy of Engineering Physics), Yibin (components, Plant 812), Guangyuan (plutonium production complex, Plant 821), Leshan |
| Textile industry | Daxian (now Dazhou), Neijiang, Suining, Nanchong |

* Although Chongqing is not counted as part of Sichuan in terms of the administrative level, it is listed here due to its geographic proximity.

Source: Gu et al. (1999: 185), Xu and Xiao (2009), Jencks (1980) and the Federation of American Scientists (FAS) (2010).

Sichuan's repurposed industrial firms later provided the necessary basis for today's local high-tech sector to flourish. In the early 2000s, with maturing electronic manufacturing capacity and skills, provincial policies started pushing for an upgrade from traditional mechanical manufacturing to a 'digital military industry/informatisation (*jungong xinxihua*)' with a focus on reforming the defence industry to adapt to information warfare, including digital security systems, AI-equipment, and combat technologies (People's Daily Online 2004). Sichuan's digital innovation industry has grown substantially, and by 2022, the high-tech industry was contributing significantly to the local GDP (Tian 2022).

The historical trajectory of Sichuan's military and defence sector helps to explain why Sichuan's high-tech industry is spread out across the province. Today, the high-tech industry clusters are not only concentrated in the provincial capital city, Chengdu, but they are also spread around in the Sichuan Basin (e.g., Chengdu, Miangyang, Meishan) as well as in mountainous areas (e.g., Guangyuan, Yibin, Ya'an, Ganzi). Many of these remote prefectures are less developed than larger urban cities in the province, but they benefitted from industrial development during the Third Front Movement (Chen 2011, 40). One leading prefecture in industrial and technological development is Mianyang prefecture, which is frequently dubbed the 'China Science and Technology City.' The prefecture is leading not only in high-tech military technologies such as AI for hypersonic weapon design but also in the commercial electronic and high-tech industries (Chen 2022).

When tracking local digital initiatives in Sichuan from 2015 to 2022 (Digital Index Database 2022), one also notices that prefecture-level digital initiatives are widely spread across the province.[4] Of the 222 initiatives tracked, in the Sichuan Basin, Mianyang prefecture topped the chart by leading 28 local digital initiatives. Chengdu city (the provincial capital) and Neijiang prefecture followed closely in second place with 27 digital initiatives. Mianyang, Chendu, and Neijiang are all prefectures that benefitted from machinery and electronic plants during the Third Front Movement (see Table 2). Guang'an prefecture is also home to many digital initiatives, likely because it is in Chongqing's spill-over zone. Overall, in Sichuan, the setup of a military defence sector during the Mao period helps to explain the regional layout and character of this province's high-tech sector.

## Zhejiang

The growth of the high-tech sector in Zhejiang has a very different origin than Sichuan's state-led development of the military complex. In this coastal province, developments have been shaped by the active role of the private sector and the relatively laissez-faire style of local governance. Today, Zhejiang is home to some of the biggest tech companies and start-ups in China. Hangzhou's Alibaba is the province's most famous tech firm, but other prominent players include Hikvision, Dataqin, Geely, NetEase, and Kuaidi Dache. In this region, the relationship between the provincial government and local private tech entrepreneurs has historically been very cooperative. Early on in their development, the provincial government became a major customer of the bigger private tech companies and provided high-tech start-ups with a certain level of freedom essential for private entrepreneurship to flourish (Breslin 2012).

Historical path dependencies play a key role in explaining the growth of Zhejiang's high-tech industry and the close public–private cooperation. Wenzhou, a prefecture in southeast Zhejiang, has long been famous as a cradle of private micro-entrepreneurs. The prefecture also played a key role in shaping the province's economic path (Tsai 2002). During the early 1980s, Wenzhou's high population density and lack of arable land pushed local government officials to promote private entrepreneurship instead of agriculture. This was very risky at the time, as private entrepreneurship in the early post-Mao period was still a political taboo and framed as the 'tail end of capitalism.' Zhejiang's local governments allowed private enterprises to formally register as public or collective enterprises, giving rise to so-called red-hat (*Hong maozi*) enterprises. These red-hat private enterprises had better access to capital and other favourable policies and, as a result, were able to grow quickly in the 1980s and early 1990s (Tsai 2002). Terms such as 'Zhejiang business culture' and the 'Wenzhou (economic) model' are still very widely used in China to describe the active bottom-up business activities that originated in Zhejiang province.

The provincial legacy of strong local entrepreneurship and its freer market environment eventually became a growth platform for the high-tech industry. Here developments were based on win-win bargains between the state and industry. Provincial leaders benefitted from a rapidly growing high-tech industry, which helped them to meet their economic growth targets in the cadre evaluation process. At the same time, the development of private tech enterprises was helped by a nurturing environment largely free of big-data technology regulation (Lv and Luo 2018). Zhejiang's provincial government also served as the main customer of tech enterprises to improve the provincial e-government services. For instance, Zhejiang was one of the first provinces to start a 'Maximum one visit for administrative procedures' digital project to showcase more efficient e-government services (Gao and Tan 2020; Kostka 2022).

Another example of close state–business cooperation is Zhejiang's 'City Brain' project. The project took shape in 2016 when Alibaba's then Chief Technology Officer Wang Jian proposed the concept to integrate different Hangzhou city administrative services in order to solve urgent city governance issues. Alibaba's City Brain project began in Hangzhou and has spread to cities throughout Zhejiang province (Chen 2021) and even abroad (Szewcow and Andrews 2020). The cooperation with Alibaba helped Zhejiang's government to position itself as the frontrunner and provincial role model for smart technologies for other provinces (14th Five-Year Plan of Zhejiang). The success of City Brain helped to deepen the provincial government's cooperation with the high-tech sector, creating new forms of mixed ownership and interdependence (Kostka 2022). In 2020, the government initiated the Zhejiang City Brain Industry Alliance, a local 'non-profit organisation,' (Zhejiang University Holding Group 2021) that comprises 331 members (as of May 2021) across the public sector, private sectors, civil organisations, and research institutions to further promote and develop

Zhejiang's flagship programme City Brain (Zhong Tuo Bang 2021). In short, for Zhejiang, the provincial hardship in the 1970s and '80s provided fertile ground for local private sector growth (Kostka 2012), which, in turn, eventually helped to produce high-tech entrepreneurs such as Jack Ma.

Zhejiang province's capital, Hangzhou, played a key role in local digital initiatives between 2015 and 2022 by offering positive spillovers to its neighbouring prefectures. The prefectures with the highest number of digital initiatives include Jinhua (53), Shaoxing (52), Huzhou (43), and Hangzhou (42), while Quzhou (16) and Taizhou (14) were significantly left behind.[5] Jinhua prefecture, the locality with the most digital initiatives, is home to dozens of industrial parks and start-up incubators and has a unique development trajectory. Jinhua's local innovation and high-tech sector was strengthened in 2015, when the Jinhua Peking University Science Park Branch was established with the help of the Jinhua Municipal Party Committee Organisation Department and Peking University (China Cultural Chamber of Commerce for the Private Sector 2017). The park's close cooperation with the prefecture's Party Committee Organisation Department and Peking University, in particular, helped to recruit top local talents from within the government and outside the park (China Cultural Chamber of Commerce for the Private Sector 2017). Shaoxing and Huzhou prefectures are geographically close to Hangzhou and benefit from positive spillover effects from Hangzhou. Alibaba's headquarters are in Hangzhou, and City Brain has been an important trademark for the entire Zhejiang province. The 'Hangzhou City Brain Experience' was repeatedly used as the benchmark for the whole province's digital initiatives work (Zhejiang Digital Economic Development Administration and Zhejiang Governance Digitalisation Promotion Committee 2020). Prefectures with the lowest number of initiatives are located farther away from Hangzhou and did not benefit from positive spillovers.

## Anhui

The agricultural province of Anhui in central China has a surprisingly large and thriving high-tech industry. In 2017, Anhui set up a fund of US$1.6 billion to support construction of the world's biggest quantum research facility (Shi-Kupfer and Ohlberg 2019, 32). The high-tech industry in Anhui is heavily concentrated in and around the provincial capital, Hefei City. Hefei is home to the Gaoxin industrial complex, which encompasses dozens of high-tech industrial parks. One of them is China Speech Valley (*Zhongguo shenggu*), which focuses on AI-powered voice recognition technologies (Hefei STIP Co. Ltd 2022). Hefei's high-tech start-ups and companies focus on integrated circuits, biomedicine, and high-end medical equipment. Within the Gaoxin industrial complex, the different industrial parks work closely together. For instance, the Gaoxin industrial complex's flagship company, iFlyTek, cooperated with another private enterprise in the Gaoxin industrial complex, ListenAI, to produce their 'AI chips' products (Su 2021).

The history of iFlyTek is the epitome of Anhui's provincial development of a local high-tech industry. In 1999, iFlytek Co., Ltd was founded in Hefei by Liu Qingfeng, then a student at the University of Science and Technology of China (USTC) in Hefei, and eighteen of his classmates. Since then, iFlytek has gradually grown into a large company that is the 'One Core (*Yī he*)' of Anhui's smart development. Among the more than one thousand companies in China Speech Valley, iFlyTek is now the largest. It has more than 14,300 employees (Market Screener 2022) and is the only intelligent speech recognition technology company listed on the Shanghai Stock Exchange (Zhu 2019: 68).

As the case of iFlyTek indicates, university linkages were key to the early start-up phases and have played an important role in attracting talent to the region. Anhui is home to USTC, a prestigious university in China that is directly managed by the Chinese Academy of Sciences. Founded in 1958 during the early years of the PRC, USTC set up many science and tech departments that are particularly relevant to emerging sciences, such as nuclear physics and space technology. The university founders include Guo Moruo, the first president of the Chinese Academy of Sciences of the PRC, who also became the first principal of USTC. Besides USTC, the Hefei University of Technology (HFUT) and Anhui University also provide a skilled labour force for the many start-up companies that have settled in China Speech Valley in Hefei.

Aside from the talent incubation and close linkages with university research institutes, Anhui's tech trajectory is also greatly influenced by its proximity to prosperous coastal neighbours. Anhui's southern prefectures enjoyed various spillover benefits on digital implementation and technological innovation from its technologically advanced neighbours. Situated along the Yangtze River, Anhui was naturally connected to the downstream cities, especially those in Jiangsu and Zhejiang. In 2014, the State Council positioned Hefei as a sub-centre city of the Yangtze River Delta city-region, which meant Hefei would be included, along with eastern coastal cities, such as Nanjing and Hangzhou, in the Yangtze River Delta's national strategy (Zhao and Zou 2018). These connections facilitated the flow of tech know-how, capital, and entrepreneurs (Kostka 2009). Table 3 summarises the ten strategic emerging industries and their regional distribution outlined in Anhui province's 14th FYP (2021–2025).

In Anhui, digital initiatives are more heavily concentrated in Southern Anhui, which is more developed than the prefectures in Northern Anhui. By tracking 205 local digital initiatives in Anhui from 2014 to 2021,[6] we find that

**Table 9.3: Anhui's strategic emerging industries in the 14th Anhui provincial FYP.**

| *District* | *Strategic Emerging Industry Clusters* | *Industry Sectors* |
|---|---|---|
| Hefei | New energy vehicles, biomedicine and high-end medical equipment, culture and creative industry, and cyber security | Smart technology/electronic appliances, AI ICTs, Green food |
| Ma'anshan | High-end computer numerical control machine tools, railway transport equipment | |
| Suzhou | Cloud computing | |
| Huainan | Big data | |
| Chizhou | Semiconductors | |
| Wuhu | Robots, new energy vehicles, modern agricultural machines, general aviation | |
| Chuzhou | Smart household electronic appliances | |
| Bengbu | Silicon-based new materials | Six new materials: bronze-based, iron-based, aluminium-based, magnesium-based, silicon-based, and bio-based |
| Tongling | Bronze-based new materials | |
| Anqing | New materials for chemical engineering | |
| Huaibei | Aluminium-based metal materials, high-end macromolecule material | |
| Huangshan | Cultural tourism | Digital culture and creative industry |
| Bozhou | Modern traditional Chinese medicine | Life and health industry |
| Fuyang | Modern medicine | |
| Xuancheng | Core basic assembly units and parts (production) | Units and parts production for machine manufacturing |
| Lu'an | High-end equipment assembly units and parts (production) | |

Source: Anhui Province 14th FYP (2021).

the leading prefectures in the south include Ma'anshan (18), Wuhu (14), and Hefei (13), while many poorer prefectures were left behind. Hefei, the provincial capital city, became the new focal point for Anhui's tech industry thanks to its industrial parks and the location of many research institutes. Ma'anshan and Wuhu are located close to both Nanjing, the capital of the prosperous Jiangsu province, and Hefei, the capital of Anhui. With more than thirty-three digital initiatives, Bozhou is an interesting exception. It is in Northern Anhui and economically in the middle among the sixteen prefecture-level cities of Anhui (Zhang 2022), but the city's economic development relies predominantly on a specific sector: traditional Chinese medicine. 'Bozhou medicine' is famous throughout China. The importance of the traditional Chinese medicine industry in Bozhou gives its digital initiatives a distinct character: the City Brain project in Bozhou, for example, was tasked with tracing and controlling the quality of traditional Chinese medicine ingredients and monitoring online vendors, along with other general functions in the area of traffic and pollution (Inspur 2022).

In summary, the high-tech industries in Sichuan, Zhejiang, and Anhui have developed differently because they have distinct historical trajectories and different levels of support for the tech sector. The historical institutional perspective offered here helps to explain why the high-tech industry in Sichuan is largely focused on civil–military industry production, while in both Anhui and Zhejiang there is a stronger focus on AI technology and speech recognition. The comparison also sheds light on the diverse state–business relations in the high-tech industry across regions: while Sichuan's tech industry has, to a large degree, been dictated by government and defence projects, Zhejiang allows more space for the tech industry to develop, albeit with tighter controls, and in Anhui, the tech industry has formed close relationships with local research institutes and labs in Hefei. Support from local governments has had a major impact on tech development in all three provinces, and historical conditions have played a formative role in shaping the outcomes.

## LOCAL REALITIES: THE LOCAL IMPLEMENTATION GAP IN HIGH-TECH POLICIES

Despite Beijing's impressive efforts to devise industrial policies for technology upgrading (Naughton 2021), there is a substantial *high-tech policy implementation gap*. The term *implementation gap* refers here to differences between Beijing's high-tech ambitions and local policy outcomes. There can be a gap because many elements of Beijing's tech agenda fall to local governments for delivery. As local governments' interests, capabilities, and pre-existing industrial structures differ widely, national plans and investment funds are often not or only partially implemented, poorly executed, or significantly delayed.

Examples of such 'gaps' include overinvestment in physical infrastructures, which causes a waste of public resources, insufficient long-term financial investments, ill-functioning digital platforms, and flawed digital services for the public. For instance, despite being home to more than 500 of the roughly 1,000 smart cities in the world (Deloitte 2018), probably less than 10 percent of the smart city projects in China were fully functional, according to a report by a Chinese think tank (Liu and Zhang 2020). Many smart cities do not offer complete services or have set up too many fragmented 'service brains' that overlap with each other, causing a waste of resources (Liu and Zhang 2020). Many digital projects face delays due to data integration and standardisation issues and fail to integrate and analyse data for predictive policies (Große-Bley and Kostka 2021). For instance, some of the widely reported local social credit pilots have so far failed to develop a functioning scoring or assessment system, while others engage only a small percentage of the entire population (Li and Kostka 2022).

One sector where high-tech failures at the local level are most apparent is the many failed cases of local government investment in the chip industry. In 2020 and 2021 alone, six Chinese multibillion-dollar chip projects filed for insolvency. Owing to the high costs and high risks involved, the semiconductor industry has become the prime example of 'an industry that is flush with state cash but still scarce on expertise' (Feng 2021).

As the next sections argue, many of the failed outcomes in local high-tech industries can partly be traced back to the insufficient financial, political, and technical capacities of the local agencies in charge of high-tech policy implementation.

## Insufficient Financial Capacities and Mismanagement of Funds

Many local high-tech projects depend on initial funding from China's Industrial Guidance Funds (IGFs). The local governments control the majority of IGFs (Naughton 2021: 109) and an estimated total sum of RMB 3.7 trillion (US$508 billion) is in the hands of prefecture governments, who are also the main implementers of the high-tech policies. In the second place are provincial governments, who control RMB 3.3 trillion (US$454 billion), while the central government controls about RMB 1.96 trillion (US$270 billion) of IGFs (Naughton 2021, 109).

At the national level, China has increased funding for technological and innovative projects (National Bureau of Statistics of China 2022) and has been planning major future investments. Local provinces have also set aside significant amounts in funding for smart city and big data projects. For instance, Guangdong province has invested RMB 10 billion in the next-generation ICT industry in Guangzhou (Liu 2017). The Guizhou provincial government has invested RMB 1 billion in a special fund for big data development that supports enterprises specialising in data collection and storage, data sharing, and information security (Wu 2021). Typically, more advanced localities in coastal provinces spend proportionally more on digital projects than less advanced localities in central and western provinces.

Despite the significant increase in funding for local IGFs, most of these tech funds are assigned to specific programmes. Local governments, whose responsibilities and tasks have skyrocketed over the past decade, tend to be seriously underfunded (Wong, 2021). As a result, costly high-tech projects are sometimes not prioritised because there are more pressing local priorities to fund. To overcome funding shortages, local governments can apply for project funding and staff expansion from the municipal, provincial, and

**Table 9.4: Examples of failed semiconductor projects**

| *City, Province* | *Local Semiconductor Company/Project* | *Type of Local Gov't Involvement* | *Year and Type of Failure* | *Reason for Failure* |
|---|---|---|---|---|
| Nanjing, Jiangsu | Tacoma/image sensor chip | In June 2016, the Nanjing Municipal Government, Nanjing Tacoma and Tal Corporation announced cooperation to build a wafer factory | July 2020, insolvency | Lack of money and dependence on foreign technology |
| Gui'an New Area, Guizhou | Huaxintong (HXT)/developing server chips based on ARM architecture | A joint venture between Qualcomm and the provincial government of Guizhou | May 2019, shut down by the government | Qualcomm headquarters shuts down server business and loses technology source |
| Haui'an, Jiangsu | Dehuai/12-inch CIS) | The Huai'an government initially attracted Joseph Lee, the chairman of Tacoma, to set up a wafer factory. After falling out with Lee, Dehuai bribed a Party member in charge of the Huai'an high-tech zone to gain project approval and construction, tax rebates and further government investment | 2020, incomplete production line construction | Other partners did not fulfil their investment obligation |

| | | | | |
|---|---|---|---|---|
| Chengdu, Sichuan | GlobalFoundries (GF)/0.18 micron/22nm chip manufacturing process | In 2016, GF and the Chongqing municipality formed a joint venture to set up a local plant | Notice of closure in May 2020 | GF cancelled investment |
| Wuhan, Hubei | Hongxin/7nm and 14nm chips | A joint venture between Wuhan's Dongxihu district (municipal) government and Beijing Guangliang Lantu Technology | 2020, insolvency | Capital chain rupture |
| Fengxi Xincheng, Xixian New Area, Shaanxi | Incoflex/flexible semiconductor | Major funding: Fengxi district development funds | 2020, insolvency | Capital shortage. Senior executives departed, leaving employees unpaid. |

Sources: Reuters (2016), Feng (2021), Chinese Semiconductor Chart/*Xin Bang* (2020), Yang (2020), Geng (2020), Li and Shi (2020).

national governments, but these funding applications are often lengthy and require sustained effort by local leadership over several years (Lo and Tang 2006; Kostka 2014).

The shortage of funding was also very clear in local governments' efforts to create a local semiconductor industry. Many of the projects described in Table 5 failed due to insufficient financial capacities. The case of Nanjing Tacoma is a vivid illustration of the local governments' limited capacity to support this costly sector. Taiwanese businessman Joseph Lee established the semiconductor company Tacoma in Taiwan in 2003. He later moved part of the business to China. In 2015, the Nanjing government invited Tacoma to the Nanjing Economic and Technological Development Zone and Tacoma signed a contract with the Israeli semiconductor giant Tower Semiconductor to buy technology know-how and IP rights from Tower for US$60 million. In 2016, Nanjing Tacoma Semiconductor Technology was officially founded. This took place against the backdrop of the 'chip rush' created by the publication of the 2014 *Guidelines to Promote National Integrated Circuit Industry Development* issued by the central government. The local government was said to have invested US$billion in Nanjing Tacoma (Zha 2016). Nanjing Tacoma was a comprehensive project that aimed to cover

**Table 9.5: Largest Local Industrial Guidance Funds**

| *Fund Name* | *Level* | *Scale (in billions of RMB)* | *Key Fund Priorities* |
|---|---|---|---|
| Optical Valley Fund (Wuhan) | Prefecture | 10 | Optoelectronic information industry, new energy and environmental protection, high-end equipment manufacturing, high-tech services |
| Kunpeng Fund (Shenzhen) | Prefecture | 50 | The upstream and downstream of the new generation of information technology industry chain |
| Shanxi Taihang Fund | Provincial | 20 | Investment in the fields of mixed ownership reform of state-owned enterprises, development zone construction, strategic emerging industries, cultural tourism industries and civil–military integration industries |
| Jiangxi Development and Upgrading Fund | Provincial | 100 | '2+6+N' industries: 2 means non-ferrous metals and electronic information; 6 means equipment, petrochemicals, building materials, textiles, food, automobiles; N means aviation, traditional Chinese medicine, mobile Internet of Things, semiconductor lighting (LED), virtual reality, energy saving, and environmental protection |
| Zhejiang Jinhua Science and Technology Park | Prefecture | 11 | Fourteen projects in the first stage. Facilities are the main priority, including necessary infrastructure for high-end education and several large industrial and innovation parks that focus on digital economy innovation and urban planning (Seetao 2021) |

| | | | |
|---|---|---|---|
| Big Data Development Fund of Guizhou | Provincial | 1 | Financially supporting the BIG DATA EXPO in Guiyang (64 million RMB), building the smart airport in the Guiyang Longdongbao International Airport, and supporting hundreds of enterprises, especially focusing on ICT |
| Guangzhou Next-Generation ICT Industry Fund | Prefecture | 10 | Building an ICT ecosystem and developing a self-sufficient ICT industry supply chain in Guangzhou city |

Sources: Naughton (2021: 109); Optical Valley Industrial Investment (2022), Shenzhen City Kunpeng Equity Investment Co., Ltd. (2022), Chin (2017), Liu (2022), Liu (2017), Wu (2021).

the whole semiconductor production chain: the plan was to set up IC design studios, R&D centres, facilities reproduction factories, assembly, testing, and packaging factories, as well as the downstream daily applications product centres. Nanjing Tacoma also promised to deliver a mass production capacity of 8-inch chips in June 2018.

It all turned out to be a big disappointment. Following a government investigation, Lee was found not to have invested any money in the project and to have relied solely on local government funds. However, according to Lee, the local government had promised him substantial support from China Invest Century Shareholding Investment Group Limited (Li and Shi 2020, 11). The park also failed to provide a credential financing guarantee company for Tacoma to lend money. Ultimately, the funding was unsustainable and far below the mark for semiconductor investments. On 19 April 2019, Lee announced that Tacoma would cease production, and eventually, the half-finished Nanjing Tacoma factory buildings were completely abandoned.

While the local government has tried to frame Tacoma's failure as the result of Lee's non-investment, the lack of sustainable government investment in the project is hard to ignore. As Tacoma's case highlights, initial state investments may be huge, but sustaining funding is a major problem. Similarly, other start-ups like Hongxin, Incoflex and Dehuai (another project started by Joseph Lee) were regarded as 'scams,' but they all followed the same pattern of starting as a high-profile project with large local government investments and high hopes from the local governments, leading to failure. The national strategic focus on semiconductor technology development developed by the central government generated an uncontrollable and wasteful 'chip rush' in the process of implementation at local levels.

## Insufficient Technical Capacities

Technical capacity constraints can further hinder the implementation of national high-tech mandates. Two pertinent technical constraints that are commonly cited in the literature are a lack of technical know-how and insufficiently trained local staff in the public sector (Segal, 2003; Fuller 2016). China's local bureaucracy is in chronic need of well-trained staff to set up and manage complex high-tech projects, which typically require tight management of outsourced subcontracts with private or state-owned enterprises. Project management for IT and high-tech initiatives can become incredibly complex. Zhang and Bao (2018) highlight the need to further enhance the high-tech literacy of leaders in key government departments and provide training to improve leaders' skills and knowledge.

China's state-owned companies generally face difficulty attracting and retaining bureaucrats with a deep understanding of technology. For example, a Chinese article that delved into the problem of a brain drain to the United States complained that young, highly trained personnel often work at state-affiliated institutions or SOEs purely for the sake of earning better *Hukou* and polishing their résumés. The article notes that one or two years after government officials get what they needed, they move to the private sector as a step toward working abroad (Lian 2022). The shortage of human capital in the tech sector also helps to explain why digital projects in China sometimes get stuck in Phase 1 (data collection), while Phases 2 (data analysis) and 3 (using data for predictive purposes) remain locked in the distant future (Große-Bley and Kostka, 2021).

## Insufficient Political and Coordination Capacities

Local governments work under certain political capacity constraints that influence policy outcomes. Political capacity constraints can result from coordination difficulties because of various factors. First, the implementation and enforcement of high-tech mandates and plans at the local level are partly hindered by bureaucratic fragmentation, as responsibilities are allocated among many different government agencies (Große-Bley and Kostka, 2021). Numerous government agencies are usually responsible for the implementation of a high-tech project but often without a clear division of labour, which in practice leads to a lack of accountability. For example, more than fifteen departments have a role to play in the implementation of digital governance platforms at sub national levels. Usually, the provincial or municipal government office and the local Big Data Administration (BDA) bureau take the lead, followed by police/legal bureaus, economic/commerce bureaus and social security/housing bureaus (Kostka, 2022). Similarly, when looking at the implementation of the City Brain projects at the local level, a lead project manager noted that there are 'way too many brains' involved (Liu and Zhang 2020). A city can have an environmental protection brain, a traffic brain, a medical care brain, and so forth—all on top of a city brain. The CEO of Hangzhou H3C City Digital Brain Research Institute, Peng Yue explained, "It might be due to the fact that many government bureaus build their own data warehouses and all call them 'brains'" (Liu and Zhang 2020). With responsibilities spread across many fragmented bureaucracies, it can be cumbersome to coordinate high-tech projects. For instance, it is often not possible for a low-ranking bureaucratic office to access relevant data from another bureau higher up the hierarchy. The same happens with bureaus at the same level. Many data systems and platforms are often only for internal use and not open to external users, even those within the government.

The implementation capacity of local departments in charge of complex high-tech projects is further constrained by competing demands and heavy workloads in implementing local agencies. High-tech projects also fall into the realm of local Development and Reform Commissions (DRCs), which are powerful but often take on too many tasks and are, therefore, sometimes understaffed. In addition, many local DRC officials lack the digital expertise to push or coordinate complex local high-tech projects. For a project to be successful, it often requires the local leadership taking it on as a pet project; only such high-priority initiatives can secure sufficient long-term start-up funding. For instance, in Zhejiang, some local digital projects have succeeded because local politicians kept pushing for them. In particular, the fast development of the 'Maximum one visit for administrative procedures' digital project would not have been successful without significant attention from Party Secretary Yuan Jiajun (Yuan, 2021).

In summary, local leaders in charge of high-tech projects receive mixed signals: they are asked to fully implement high-tech projects, but these demands by upper-level governments are not always matched by a corresponding increase in political power and financial resources. The following quotation summarises the challenge quite well: 'In a word, Big Data bureaus are a paradise for innovators, but hell for those who follow prescribed routines. Work in Big Data bureaus can be summarised with the following key words: endless tasks, limited budget compared with other departments, glory, outstanding performance, bright future' (Zhang, 2020).

## CONCLUSION

This chapter argued that China's technological rise is both fueled and constrained by the specific institutions of the party-state. The 'fuel' is the party-state's capacity and will to lead China up the value chain thanks to massive investments. China's policymakers have shifted from an indicative planning approach to new industrial policies in which the state plans to invest unprecedented sums to leapfrog in technology—thereby helping Chinese tech firms to advance rapidly in the global tech war.

The 'constraints' involve the downside of decentralised and fragmented authoritarianism. As illustrated with the cases of Anhui, Sichuan, and Zhejiang provinces, at the provincial level, the trajectory of technology industrial growth differs across regions and is often shaped by multiple pre-existing economic, social, and political factors. Furthermore, many national tech ambitions are often only partially implemented at the local level due to local governments' insufficient technical, financial, and political capacities to push for tech leadership in their locally grown tech industries. In particular, the lack of long-term finances has been the main hurdle to the development of viable high-tech industries at the local level.

## NOTES

1. The author gratefully acknowledges funding the European Research Council (ERC Starting Grant No: 852169). The author is also very grateful for excellent research support from Jingshin (Anita) Lin.

2. Despite strong ambition, the funds ultimately raised only about US$672 billion in total (Luong et al. 2021, 4).

3. The seven industries are: cloud computing, Big Data, Internet of Things, Industrial Internet, Blockchain, artificial intelligence and virtual and augmented reality. The ten sectors are: smart transport, smart energy, smart manufacturing, smart agriculture and irrigation, smart education, smart health, smart culture and tourism, smart community, smart household, and smart government.

4. The spread of the 222 digital initiatives in Sichuan is as follows: Mianyang (28), Chengdu (27), Neijiang (27), Guang'an (22), Meishan (13), Ganzi (12), Dazhou (12), Bazhong (10), Suining (9), Nanchong (9), Ya'an (8), Luzhou (8), Guangyuan (7), Deyang (7), Ziyang (6), Aba (5), Leshan (3), Panzhihua (3), Yibin (3), Zigong (2) and Liangshan (1).

5. The spread of the 365 digital initiatives in Zhejiang is as follows: Jinhua (53), Shaoxing (52), Huzhou (43), Hangzhou (42), Lishui (32), Ningbo (30), Jiaxing (23), Zhoushan (21), Wenzhou (20), Quzhou (16), Taizhou (14).

6. The distribution of the 205 digital initiatives in Anhui is as follows: Bozhou (33), Chizhou (22), Maanshan (18), Huangshan (18), Wuhu (14), Lu'an (13), Hefei (13), Chuzhou (11), Fuyang (10), Huaibei (10), Xuancheng (9), Benggu (8), Tongling (8), Anqing (8), Suzhou (6) and Huainan (4).

## REFERENCES


Abbany, Z. 2017. Tianhe-3: China building exascale supercomputer. DW. 20 February. Available from: https://www.dw.com/en/tianhe-3-china-says-its-world-first-exascale-supercomputer-will-be-ready-by-2020/a-37635237 [11 November 2022].

Anhui Province 14th Five-Year Plan. Anhui Provincial Standing Committee. 2021. Anhui sheng guomin jingji he shehui fazhan di shisi ge wu nian guihua he 2035 nian yuanjing mubiao gangyao, issued February. Available from: https://www.ndrc.gov.cn/fggz/fzzlgh/dffzgh/202104/P020210408558077118783.pdf [16 November 2022].

Breslin, S. 2012. Government–Industry Relations in China: A Review of the Art of the State. In A. Walter, and Xiaoke Zhang, ed., *East Asian Capitalism: Diversity, Continuity, and Change.* Oxford: Oxford Academic, 29–45.

Bu, Rixin. 2020. Zhongguo xinxi chanye que xin shao hun, zhe yi hun ruhe jiejue? (China's ICT industry is lacking core and soul, how should this core be earned?). *eefocus*. 19 August. Available from: https://www.eefocus.com/mcu-dsp/471240 [7 June 2022].

Butterfield, F. 1980. China's Military Industry Pursues Civilian Markets. *New York Times*. 1 June. Available from: https://timesmachine.nytimes.com/timesmachine/1980/06/01/114114707.html?pageNumber=8 [6 June 2022].

Chen, Liping. 2021. Zhejiang sheng chengshi danao chanye lianmeng chengli, chengshi danao 2021 nian jiang fugai quansheng (Zhejiang province City Brain Industry Alliance founded, City Brain is going to cover the whole province in 2021). *People's Daily Online*. 06 January. Available from: http://zj.people.com.cn/n2/2021/0106/c186327-34513244.html [4 October 2022].

Chen, Minglu. 2011. Mianyang: The Legacy of State Socialism and Local Construction. *Provincial China* 3(1), 34–59.

Chen, Stephen. 2022. AI on its way to replacing humans in hypersonic weapon design: Chinese study. *South China Morning Post*. 23 March. https://www.scmp.com/news/china/science/article/3171533/ai-its-way-replacing-humans-hypersonic-weapon-design-chinese [2 July 2022].

Chin, Hong. 2017. Jiangxi sheli shou zhi qian yi ji fazhan shengji yindao jijin cu fazhan (Jiangxi set up the first hundred-billion-level development upgrade fund to boost development). *Xinhuanet.com*. 25 February. Available from: http://www.gov.cn/xinwen/2017-02/25/content_5170937.htm [10 November 2022].

China Cultural Chamber of Commerce for the Private Sector. 2017. The practice and exploration of the innovation of the party organization in Jinhua Zhejiang Science. *Chinaci.org*. 9 March. Available from: https://www.chinaci.org/wap/en/commentanalysis/1859.html [2 November 2022].

Chinese Semiconductor Chart/ Xin Bang. 2020. Daobi, xie ben wu gui! 400 yi Kun Tong Bandaoti: Huang liao! (Bankrupt! No return of the investments! 40 billion RMB Incoflex Semicondcutor: Doomed!). *WeChat News.* 27 April. Available from: https://mp.weixin.qq.com/s?__biz=MzI1NjIyODU4Ng==&mid=2247484271&idx=1&sn=bc9f8001ab6244c8f20ec800b5248961&scene=45 [31 October 2022].

Corbett, T. and Singer, P. 2022. China may have just taken the lead in the quantum computing race. *Defence One*. 14 April. Available from: https://www.defenceone.com/ideas/2022/04/china-may-have-just-taken-lead-quantum-computing-race/365707/ [6 October 2022].

Deloitte. 2018. Supercharging the Smart City—Smarter people and better governance. Deloitte Perspective Volume VII. *Deloitte*. Available from: https://www2.deloitte.com/cn/en/pages/about-deloitte/articles/deloitte-perspective-v7-chapter7.html [6 June 2022].

Disha. 2021. Quantum Computing: Top Countries Participating in Quantum Race. *Global Tech Outlook*. 16 August. Available from: https://www.globaltechoutlook.com/quantum-computing-top-countries-participating-in-quantum-race/ [20 October 2022].

Federation of American Scientists. 2010. Subtitled Clips of Chin"s Declassified Underground Nuclear Facility in Chongqing. DNI Open Source Center. 23 April. Available from: https://nuke.fas.org/guide/china/facility/chongqing.pdf [6 October 2022].

Feng, Emily. 2021. A cautionary tale for China's ambitious chipmakers. *NPR*. 25 March. Available from: https://www.npr.org/2021/03/25/980305760/a-cautionary-tale-for-chinas-ambitious-chipmakers [31 October 2022].

Fuller, Douglas. 2016. Paper tigers, hidden dragons: Firms and the political economy of China's technological development. Oxford University Press.

Gao, Xiang and Tan, Jie. 2020. From web to Weber: Understanding the case of "One-Go at Most" as ICT-driven government reform in contemporary China. *China Review* 20(3), 71–97. Available from: https://www.muse.jhu.edu/article/764071.

Geng, Yvonne. 2020. GlobalFoundries abandons Chengdu wafer fab. *EETimes*. 26 May. Available from: https://www.eetimes.com/globalfoundries-abandons-chengdu-wafer-fab/ [31 October 2022].

Große-Bley, J. and Kostka, G. 2021. Big Data dreams and reality in Shenzhen: An investigation of Smart City implementation in China. *Big Data & Society* 8(2). Available from: https://doi.org/10.1177/20539517211045171

Gu, Chaolin, Tsai, Jianming, Niou, Yafei, Suen, Ying, Chen, Tian, Chai, Yenwei and Ye, Jiaan. 1999. *Zhongguo chengshi dili* (Chinese Cities Geography), Shanghai: Commercial Press.

Hefei STIP Co. Ltd. 2022. Chanye yuanqu (Catalog of the industrial parks). *Hefei STIP Co. Ltd.* Available from: https://pinyin.thl.tw/converter?s=□□□□&c=title [7 November 2022].

Huang, Xin. 2016. Tuidong Zhongguo zhizao xiang Zhongguo chuangzao zhuanbian. (Pushing transformation from Made in China to Invented in China). *Economic Daily*. 30 August. Available from: http://www.gov.cn/xinwen/2016-08/30/content_5103373.htm [13 October 2022].

Inspur. 2022. Bozhou 'chengshi danao,' yong dashuju shouhu chengshi wendu (Bozhou City Brain: Protect the warmthof thee city with the help of big data). *Inspur*. Available from: https://www.inspur.com/lcjtww/2315750/2320981/2345517/2345521/2548085/index.html [9 November 2022].

Isaksen, A. 2015. Industrial development in thin regions: trapped in path extension? *Journal of Economic Geography* 15(3), 585–600. https://doi.org/10.1093/jeg/lbu026.

Jencks, H. W. 1980. The Chinese "Military-Industrial Complex" and Defence Modernization. *Asian Survey* 20(10), 965–89.

Kania E.B. and Laskai L. 2021. Myths and Realities of China's Military-Civil Fusion Strategy. *Center for a New American Security (CNAS)*, 28 January. Available from: https://www.cnas.org/publications/reports/myths-and-realities-of-chinas-military-civil-fusion-strategy [9 September 2022].

Kharpal, A. 2019. With Xi's backing, China looks to become a world leader in blockchain as US policy is absent. *CNBC*. 15 December. Available from: https://www.cnbc.com/2019/12/16/china-looks-to-become-blockchain-world-leader-with-xi-jinping-backing.html [19 October 2022].

Kostka, G. 2022. Digital Governance in China. *SSRN Working Paper*. Available from: https://papers.ssrn.com/sol3/papers.cfm?abstract_id=4047764

Kostka, G. 2014. Barriers to the implementation of environmental policies at the local level in China, World Bank Policy Research Working Paper no. WPS 7016. Washington, DC: World Bank Group. Available at: http://documents.worldbank.org/curated/en/2014/08/20144757/barriers-implementation-environmental-policies-local-level-china [10 November 2022].

Kostka, G. 2012. Mobility and agency: Private sector development in rural central China. *The China Journal* 67, 47–66. https://doi.org/10.1086/665739

Kostka, G. 2009. Private sector development in Anhui Province: The impact of regional spillovers from Jiangsu Province. In Zhongmin Wu, ed., *China in the World Economy*, London: Routledge, 212–34.

Landry, P.F. 2008. *Decentralized Authoritarianism in China: The Communist Party's Control of Local Elites in the Post-Mao Era.* Cambridge: Cambridge University Press.

Lee, Dongmin. 2011. Swords to ploughshares: China's defence conversion policy. *Defence Studies* 11(1), 1–23. doi: 10.1080/14702436.2011.553101.

Li, Haili and Kostka, G. 2022. Accepting but not engaging with it: Digital participation in local government-run social credit systems in China. *Policy & Internet*, 1–30.

Li, Jiaxing. 2021. Zhongguo gaoxiao weishenme yao jianli weilai jishu xueyuan? (Why would Chinese high-education institutions set up the School of Future Technology?) *The Initium*. 2 June. Available from: https://theinitium.com/article/20210602-technology-future-schools/ [2 July 2022].

Li, Jiaxing. 2021. Zhongguo gaoxiao weishenme yao jianli weilai jishu xueyuan (Why do Chinese higher-educations found the School of Future Technology?). *The Initium.* 2 June. Available from: https://pinyin.thl.tw/converter?s=中國高校為什麼要建立未來技術學院□&c=lower [7 June 2022].

Li, Lianjiang and O'Brien K. J. 1999. Selective policy implementation in rural China. *Comparative Politics* 31(2), 167–86.

Li, Xiaoguang and Shi, Dan. 2020. Zhongguo xinpian chanye 'lan wei.' diaocha: Shei wie chongdong maidan? (Investigation into the Chinese chip industry 'rotten tails': Who pays for the blind impulse?). *Business School*. November, 9–15. Available from: https://finance.sina.com.cn/china/2020-11-03/doc-iiznctkc9234057.shtml [31 October 2022].

Lian, Si. 2022. Woguo gaokeji rencai peiyang lujing tanxi (Study on the route of our country's high-tech talents cultivation). *People's Forum*. 26 May. Available from http://www.rmlt.com.cn/2022/0526/648044.shtml [1 November 2022].

Liu, Jingfeng and Zhang, Heng. 2020. 7 nian le, weisheme zhihui chengshi hai bu zhihui? | Jiazi Guangnian (7 years have passed, why aren't Smart Cities smart? | Jiazi Guangnian) *Jiazi Guangnian*. 15 December. Available from: https://mp.weixin.qq.com/s/YHDkiJgyI-b3kmUcWFzTlg [15 August 2022].

Liu, Yi. 2022. Jiangxi 600 yi yindao jijin, lai le (Jiangxi's 60 billion government guiding fund is coming). *Sina Technology.* 17 May. Available from: https://finance.sina.com.cn/tech/2022-05-17/doc-imcwipik0351234.shtml [10 November 2022].

Liu, Xin. 2017. [Guangzhou ribao] 'Liang ge huigui' shengeng bentu, Guangzhou jijin fadong '3·50' jihua ([Guangzhou Daily] 'The two comebacks' deeply engaged locally, Guangzhou fund activated the '3.50' plan). *Guangzhou Daily*. 29 December. Available from: https://www.sfund.com/news/shownews.php?id=301 [11 November 2022].

Lo, Wing-Hung C. and Tang, Shui-Yang. 2006. Institutional reform, economic changes, and local environmental management in China: The case of Guangdong Province. *Environmental Politics* 15(2), 190–210.

Luong N., Arnold Z. and Murphy B. 2021. Understanding Chinese government guidance funds: An analysis of Chinese-Language sources. *Center for Security and Emerging Technology (CSET)*, March. Available from: https://cset.georgetown.edu/publication/understanding-chinese-government-guidance-funds/ [21 October 2022].

Lv, Aofei and Luo, Ting. 2018. Asymmetrical power between internet Giants and users in China. *International Journal of Communication* 12, 3877–95.

Market Screener. 2022. IFLYTEK Co. Ltd. *Market Screener*. Available from: https://www.marketscreener.com/quote/stock/IFLYTEK-CO-LTD-6500190/ [19 October 2022].

Naughton, B. 2021. *The Rise of China's Industrial Policy 1978 to 2020* (Primera edición). Universidad Nacional Autónomica de México, Facultad de Economía.

National Development and Reform Commission (NDRC). 2016. The 13th Five-Year Plan for economic and social development of the People's Republic of China 2016–2020, Central Compilation & Translation Press of PRC, issued December. Available from: https://en.ndrc.gov.cn/policies/202105/P020210527785800103339.pdf [4 November 2022].

National Development and Reform Commission (NDRC). 2021. Zhonghua renmin gongheguo guomin jingji he shehui fazhan di shi si ge wu nian guihua he 2035 nian yuanjing mubiao gangyao (The 14th Five-Year Plan of the PRC), issued 23 March. Available from: https://www.ndrc.gov.cn/xxgk/zcfb/ghwb/202103/t20210323_1270124.html?code=&state=123 [4 June 2022].

National Bureau of Statistics of China. 2022. China's R&D expenditure reached 2.79 trillion Yuan in 2021, issued January. Available from: http://www.stats.gov.cn/english/PressRelease/202201/t20220127_1827065.html [24 October 2022].

Olsen, Sam. 2020. China is winning the war for global tech dominance https://thehill.com/opinion/technology/518773-china-is-winning-the-war-for-global-tech-dominance/ [24 October 2022].

Optical Valley Industrial Investment. 2022. Business area. *Optical Valley Industrial Investment*. Available from: http://www.whovii.com/public/portal/list/index/id/4.html [10 November 2022].

People's Daily Online. 2004. China's military manufacturing industry heading for informationization and digitization. *People's Daily Online*. 26 March. Available from: http://en.people.cn/200403/26/eng20040326_138573.shtml [6 June 2022].

Reuters. 2016. Qualcomm unveils $280 million joint venture with Chinese province. *Reuters*. 17 January. Available from: https://www.reuters.com/article/us-qualcomm-china-jointventure-idUSKCN0UV10A [31 October 2022].

Segal, A. 2003. Segal, A. (2003). Digital dragon: high-technology enterprises in China. Cornell University Press.

Segal, A. 2018. When China rules the web: Technology in service of the state. *Foreign Affairs* 97(5), 10–8.

Shenzhen City Kunpeng Equity Investment Co., Ltd. 2022. Company Profile. *Shenzhen City Kunpeng Equity Investment Co., Ltd.* Available from: http://www.kpcapital.cn/about.aspx?TypeId=1&FId=t1:1:1 [10 November 2022].

Shi-Kupfer, K. and Ohlberg, M. 2019. China's digital rise: Challenges for Europe. *Mercator Institute for China Studies (MERICS)*, 8 April. Available from: https://www.merics.org/sites/default/files/2020-06/MPOC_No.7_ChinasDigitalRise_web_final_2.pdf [13 October 2022].

Su, Jianxun. 2021. Ke Da Xun Fei tuijin AIoT zhanlüe, yu Ling Si Zhineng lianhe tuichu xinpian jiejue fangan (iFlyTek marched forward with AIoT strategy, launching chips solution with Listenai). *36Krchuhai*. 29 March. Available from: https://letschuhai.com/kedaxunfeituijin-aiot-zhanlueyulingsizhinenglianhetuichuxinpianjiejuefangan [6 July 2022].

Szewcow, B. and Andrews, J. 2020. Kuala Lumpur to build 'City Brain' with Alibaba Cloud. *The UN specialized agency for ICTs*. 7 April. Available from: https://www.itu.int/hub/2020/04/kuala-lumpur-to-build-city-brain-with-alibaba-cloud/ [6 July 2022].

Tian, Jiao. 2022. Sichuan's high-tech industry's revenue in the first half of the year exceeds one trillion. *Inf News*. 10 November. Available from: https://inf.news/ne/economy/37bfdf79a3f9513c7ead76232b2b6317.html [10 November 2022].

Tsai, Kellee. 2002. *Back-Alley Banking: Private Entrepreneurs in China,* Ithaca, New York: Cornell University Press.

Webster G., Creemers R., Kania E. and Triolo P. 2017. Full translation: China's 'New generation artificial intelligence development plan.' *DigiChina.* 1 August. Available from: https://digichina.stanford.edu/work/full-translation-chinas-new-generation-artificial-intelligence-development-plan-2017/ [13 October 2022]

Wong, Christine. 2021. Plus ça change: Three decades of fiscal policy and central-local relations in China. *China: An International Journal, 19(4),* 1–31.

World Intellectual Property Organization (WIPO). 2022. Global Innovation Index 2021—China. Available from: https://www.wipo.int/edocs/pubdocs/en/wipo_pub_gii_2021/cn.pdf [8 October 2022].

Wu, Hua. 2021. 70 yi yuan! Guizhou shengji caizheng sheli xinxing gongyehua fazhan jijin (70 billion RMB! Guizhou provincial finance set up new industrialization development fund). *Guiyang Daily.* 13 July. Available from: http://www.gywb.cn/system/2021/07/13/031407975.shtml [11 November 2022].

Xi, Jinping. 2022. Buduan zuoqiang zuoyou zuoda woguo shuzi jingji (Continuously strengthening, optimizing, growing our country's digital economy). *Qiushi.* 15 January. Available from: http://www.qstheory.cn/dukan/qs/2022-01/15/c_1128261632.htm [20 October 2022].

Xu, Bin and Xiao, Linxing. 2009. Planning and construction history of Panzhihua during the three-front strategy period: Backgrounds, process, and mechanism. Paper presented at the 15th International Historical Planning Society, Sao Paolo, Brazil. 15–18 July. Available from: http://www.usp.br/fau/iphs/abstractsAndPapersFiles/Sessions/31/XU_XIAO.PDF [29 September 2022].

Yang, Sophia. 2020. Six of China's largest semiconductor projects now halted. *Taiwan News.* 5 October. Available from: https://www.taiwannews.com.tw/en/news/4023601 [31 October 2022].

Zha, Jingzhong. 2016. Dekema bandaoti chanyeyuan kaigong jianshe (Nanjing Tacoma Semiconductor Technology industrial park construction laying foundation). *Nanjing Daily.* 9 June. Available from: https://news.sina.cn/2016-06-09/detail-ifxszfak3441621.d.html [8 November 2022].

Zhang. 2022. Bozhou shi jingji zongliang jinru quan sheng qian ba (Bozhou city's total economic performance ranking in 8th place in the province). *Anhui News.* 3 November. Available from: http://ah.anhuinews.com/bozhou/news/jj/202211/t20221103_6488121.html [9 November 2022].

Zhang, Xiao and Bao, Jing. 2018. Shuzi zhengfu ji pingtai: Yingguo zhengfu shuzi hua zhuanxing zhanlüe yanjiu ji qi qishi (Digital government as a platform: a study of the uk government's digital transformation strategy and its implications). *Chinese Public Administration* (18)3, 27–32.

Zhao, Wanxia and Zou, Yonghua. 2018. Hefei: An emerging city in inland China. *Cities* 77, 158–169. Doi: https://doi.org/10.1016/j.cities.2018.01.008

Zhejiang Digital Economic Development Administration and Zhejiang Governance Digitalisation Promotion Committee. 2020. Guanyu yinfa shenru guanche Xi Jinping zongshuii kaocha Zhejiang zhongyao jianghua jingshen jiakuai chengshi danao jianshe yu tuiguang gongzuo fangan de tongzhi (Notice on profoundly implement the spirit of General Secretary Xi Jinping's important speech in his official inspection in Zhejiang), issued 11 August. Available from: https://zjjcms-public.oss-cn-hangzhou-zwynet-d01-a.internet.cloud.zj.gov.cn/jcms_files/jcms1/web3241/site/attach/0/6201073d6bfb41e696e4090ac8a77662.pdf [2 November 2022].

Zhejiang University Holding Group. 2021. Zhejiang chengli chengshi danao chanye lianmeng, Tongdun keji rong ren shou pi huiyuan danwei (Zhejiang formed the City Brain Industry Alliance, Tongdun Technology honourly elected in the first group of members). *Tongdun Technology*. Available from: http://www.kggs.zju.edu.cn/index.php?a=detail&id=2261 [7 October 2022].

Zhongguo Jiaoyu Zaixian (eol.cn). 2021. Shier suo 'yiliu daxue' ruxuan! Jiaoyubu gongbu shoupi weilai jishu xueyuan mingdan (12 'excellent' colleges elected! Ministry of Education announced the list of the first round of the School of Future Technology') 26 May. Available from: https://www.eol.cn/news/yaowen/202105/t20210526_2113699.shtml [6 June 2022].

Zhong Tuo Bang. 2021. [Yitou xiangmu dongtai] Shuqin keji chengwei Zhejiang sheng chengshi danao chanye lianmeng chengyuan danwei ([News on invested items] Dataqin became a member of the Zhejiang City Brain industry alliance). *Phoenix News*. 31 May. Available from: https://ishare.ifeng.com/c/s/v002D1k1M-HYDfFRr9VI--j6HYGCRjV9qEUHKunwkKB7VLVmE [21 October 2022].

Zhu, Xiaoming. 2019. *Emerging Champions in the Digital Economy: New Theories and Cases on Evolving Technologies and Business Models*, Berlin/Heidelberg: Springer.

Zou, Shuo. 2021. 'Future technology schools' get green light. *China Daily*. 5 June. Available from: https://www.chinadaily.com.cn/a/202106/05/WS60bab74ba31024ad0bac3c42.html [7 November 2022].

*Chapter 10*

# Opening the City Through Debordering IT

## *The Making of an Innovation Ecosystem in a Post-Industrial Special Economic Zone in China*

Yujing Tan

During Chinese economic reform, the Shenzhen Special Economic Zone (SEZ) has been envisaged as a site of experimentation to exhibit the success of the Socialist market economy. The 2018 urban planning agenda regarding Innovation and smart cities turned the Shenzhen SEZ into an open city by removing the borders between its inner and peripheral city and constructing infrastructure for individualistic IT-driven enterprises, aiming to build an "international smart city" (Jiang and Xuan 2018). This chapter uses the development of the Shenzhen SEZ as a case-study to examine the geopolitical and social implications of how local government and enterprises, as economic stakeholders, responded to these industrial and urban planning policies. It asks which mechanisms these stakeholders used to transform the Shenzhen SEZ into an innovation hub, providing an on-the-ground account of how smart state and smart city policies are implemented. Through this analysis, the chapter hopes to contribute to a better understanding of how communities implement industrial and urban planning policies more broadly.

Data for this study was collected through interviews and participant observations from 2015 to 2017 and was supplemented by video interviews in 2021 and 2022 as well as official documents, historical materials, and newspaper articles. I interviewed official local government planners, officials working in tech-innovation and entrepreneurship, forerunners and current immigrants in Shenzhen, and tech-entrepreneurs-to-be running IT startups.

This chapter is structured as follows. First, the chapter traces how an entrepreneurial borderland was formed in Shenzhen through informal economic movement after the introduction of market reforms (1978). Second, it explains how state-led urban and industrial planning created a market transition, new government-business relations, and new human-capital mobility in Shenzhen. I attempt to argue that this planning of a city of innovation and smart city is gradually formalising informal economic forces and repositioning Shenzhen, the former "world factory," in the global supply chain. This transformation has made Shenzhen a model for innovation in China. Third, the chapter will conclude with a conceptual reflection on the politics of techno-spatial planning in Chinese entrepreneurial borderland of innovation.

## THE FORMATION OF AN INNOVATION BORDERLAND: INFORMAL ECONOMY IN SOUTH CHINA'S SPECIAL ECONOMIC ZONE

Shenzhen has been subject to the Chinese state's industrial transformation and economic policy reform since the 1980s. But what do these transformations and reforms really mean for our understanding of innovation in China? This section will provide an overview and analysis of how reform and opening-up policies played out in Shenzhen. The city was assigned the role of a special economic zone and innovation hub. In this way, the state sought to encourage economic dynamism, private entrepreneurship, and the mobility of the domestic population. The relationship between the state and the market in Shenzhen, especially between the local government and private enterprises, was also reshaped in this process. A brief historical analysis of the Shenzhen SEZ will help illuminate how innovative enterprises, and innovation in China more generally, formed and accumulated in this borderland.

This section highlights how the transformation that has shaped modernity and mobility in this borderland was driven by two imaginary worlds: the less developed mainland China, on the one hand, and the developed "west," on the other. This outdated binary was cultivated by the oligarchy-driven cold war but also by unbalanced regional economic development: compared to the capitalist world, the socialist world is poor. The two-world binary was strengthened by China's economic reforms in 1978. In order to revive the stagnant domestic economy and its developmental project in the world economy, the Chinese central state imitated the "Four Asian Tigers" (Hong Kong, Singapore, South Korea, and Taiwan) in the 1970s. In 1979, the state picked four coastal locations (Shenzhen, Zhuhai, Shantou, and Xiamen), administratively promoted them to cities, and entitled them SEZs in order to draw foreign export-oriented and labor-intensive manufacturers away from Hong Kong. Capital and labor flowed into Shenzhen. The first Chinese stock exchange center was built there in 1990. The introduction of (neo-)liberal stateless financial regulation in Shenzhen successfully moved foreign capital to mainland China.

The earliest Shenzhen SEZ was separated by two borders. The metropolitan side of this second border-line is the SEZ, which has shared a national (first) border-line with the United Kingdom since 1898, while the other side out of the second border-line is made up of "factory zones" established for foreign-invested manufacturing enterprises since the 1980s. Local people called the metropolitan side of the second border-line the inner area (*guannei*) and the other side, on which many immigrant workers lived, the outside of the border (*guanwai*). Since 1980, the factory zone in *guanwai* has attracted a large amount of foreign capital and many immigrants from within China, especially rural areas. Foxconn, the largest electronic manufacturing contractor in the world, is also located in this huge factory cluster. In addition to capital from Taiwan and Hong Kong, Japanese and Korean high-tech companies such as Panasonic, Sony, and Samsung invested during this period, outsourcing their production lines to private manufacturers in Shenzhen with the support of local government policies. Thousands of Three Import and Compensation Trade Enterprises[1] were established in the 1990s. This influx of foreign capital led to an influx of immigrants from mainland China (especially from Hunan and Jiangxi, provinces neighboring Guangdong). Many immigrants even quit permanent contract jobs in their hometowns to come to Shenzhen.[2] The last years of the 1990s witnessed declining border control. In 2003, the national Regulation on Custody and Repatriation, which strictly limits domestic migration without official permits, was relaxed. Shenzhen's government started to deconstruct the second borderline and loosen its control over the mobility of the population.

The Shenzhen that is described as an economic miracle—in mass media, in official brochures, and among developers—actually arose from increased mobility under state-sponsored economic liberalisation and a sustainable informal urban economy. It is the informal urban economy brought by immigration and entrepreneurship that sustained the imbalanced development of China's political economy: economic liberalisation without political liberalisation. Counterintuitively, in the context of Shenzhen's export-oriented industrialisation, the informal economy was supported by the state's regulation of economic development. In order to develop the local economy, the local state chose to turn a blind eye to the poor welfare system workers were offered by the private sector. Indeed, the state offered no public services (e.g., infrastructure and security regarding labour rights) during the early urbanisation period and export-oriented industrialisation of the 1980s and '90s. Drawing data from Shenzhen and other cities in the Pearl River Delta, many sociologists assert that most private sectors in China are dominated by informal economy (Hu and Zhao 2006; Huang 2009).

During the central states retreat and economic liberalisation in the 1980s, Shenzhen became an enclave to test and implement national economic policies and spatial planning practices imitated from developed countries. This is not to say that the central state completely retreated during the market transition. On the contrary, by marketising central state-owned sectors (marketising the Merchant Bureau into the China Merchants Group, for example) and state enterprises in Shenzhen, the Chinese state strengthened its economic power in the market economy (Yang 2001; Pieke 2009). At the same time, even as the Tax-sharing Reform consolidated the central state's economic authority in 1994, local government in Shenzhen still gained agency in local economic reforms. It is this institutional dynamic that drove mobility in Shenzhen: the state-sponsored economic liberalisation attracted human and global capital. However, mobility implies not only to the inflow of human capital, but also to its outflow. Recent deindustrialisation, moving away from labour-intensive manufacturing and toward the technology industry, led to an outflow of factory workers and an inflow of IT literate individuals: enter the young professionals (Wang and Tan 2020).

The local state-led urbanisation initiated in 2003 reshaped Shenzhen into the "first Chinese city without villages and villagers" and "a modernised and internationalised city" (Pu and Li 2003). Shenzhen has been re-framed into a smart city and city of innovation since 2011 (Shenzhen Government 2012; Shenzhen STITITC 2013). The urban renewal policy that did this further promotes the mobility of land as a form of capital. Former urban villages were gradually transformed into real estate companies limited by shares. Villagers ceded land use rights to local governments and became shareholders. On this land, the local government quickly established new technology parks to support companies such as Tencent and Huawei and internationalise them into the global market. Obsolete factory zones were gentrified into creative clusters designed in the images of Greenwich Village in New York and co-working spaces in Silicon Valley.[3]

Local government's industrial policy practices have reshaped its relationship with enterprises. Foreign capital and export-oriented enterprises, which were supported by policies at the beginning of the reform and opening-up period, gave way to technology companies representing "independent innovation" in important strategic sectors (e.g., biotechnology, the internet, the cultural and creative industry, new energy, new materials, and new generations of information technology). In 2015, the second borderline had been wholly deconstructed. "There is no need to set the segregated line anymore, because there is no sharp imbalance in economic development between the inner metropolitan city of Shenzhen and the peripheral city of Shenzhen,"[4] a retired local official told me. In addition, Shenzhen has become the driver of new industrial regionalisation and globalisation. The large, cheap electronics production chains are leaving the peripheral city of Shenzhen and being relocated to Southeast Asia and other western and northern provinces (e.g., Guizhou and Hebei) in mainland China. The central urban districts are being highly gentrified by real estate tycoons, with expensive residential communities, hi-tech parks, financial centers, and large shopping malls.

## THE MAKING OF AN INNOVATION ECOSYSTEM: FORMALISATION OF THE INFORMAL ECONOMY

Under the aegis of a state project to form an innovation hub in Shenzhen, IT-focused startups are burgeoning. Private, informal economic forces are alive in the fashioning of Shenzhen into a smart city and city of innovation. However, as part of the political economy of China's economic reforms, building information-friendly cities entails formalising informal urban economies. Here, "formalisation" does not mean that robust informal economic forces are weakened and suppressed. Rather, it means that the state actively joins in the production of new informal economic forces by building "innovative" and "smart" urban infrastructure. It is the government that defines the boundary of the formal and the informal in the market economy.

### Upgrading SEZ: Markets and Talents in the Transition from "Copycat-China" to "Innovative China"

In Shenzhen, making a city of innovation and smart city drives an upgrading project that excludes the less upgraded export-oriented manufacturers and propels flexible startups into the global supply chain of smart goods. This chapter argues that Shenzhen's upgrading is intertwined with the city's transformation into a simultaneously marketplace, which further promotes the accumulation and stratification of talent and capital in innovation-based industries.

Shenzhen's Huaqiangbei electronics market is a prime example of this shift. Often tagged as the biggest market of copycat electronics in the world, Huaqiangbei electronics market has attracted merchants from all over. Since the 2010s, the central government has started trying to dispel the negative image of Chinese products, launching a series of policies to punish producers of counterfeit goods and passing laws on issues of intellectual property (State Council 2010). Following the central government's crackdown on counterfeit goods and intellectual property infringement, local government launched a purge of Huaqiangbei electronics market (Ifeng Finance 2013).[5] An informant working in the market, Cui, still remembered the days when "a lot of merchants had to quickly destroy the fake goods. Otherwise they would be harshly punished by the troops to crack down on counterfeit goods (*dajiadui* 打假队)."[6] Within several years, the market was no longer a hub for counterfeit goods. Huaqiangbei market is now seen as potentially the biggest consumer-end products market in the world. Its foreign customers are not only big company buyers of electronic components, but also individual professionals and opportunity seekers who legally or illegally affiliate themselves with local workplaces and start their own businesses in Shenzhen. In other words, more and more individual foreigners (not only company expats) are becoming producers, joining in the production of Chinese brands and goods to meet the demands of broadening domestic markets.

In addition, China's market shift has produced a stratified class of foreigners in Shenzhen. An informant, Xiao Ling, grasped this situation. The manager of a small store who runs both retail and wholesale electronic kits businesses in Huaqiangbe, Xiao explained that a few years ago most of the foreigners walking through the market were Middle Easterners and Africans. Now, however, their biggest buyers come from developed countries; often they are Americans or Europeans working near Huaqiangbei. They come to buy equipment to work with their Chinese business partners in the nearby incubation centre. It seems that the socio-economic stratification of foreigners in the Chinese market is accelerated by its shift. The shift from an export-oriented manufacturing economy to a consumer-driven innovation economy asks for professional workers with diverse professional backgrounds to answer the sophisticated demands of both domestic markets and overseas markets. This prepares the market and talent for the building of China's innovation ecosystem.

In line with the state's reshaping of the market, local government also encourages young IT professionals to start their own businesses in the technology sector under the guidance of the Mass Entrepreneurship and Innovation policy (Stace Council 2015).[7] This policy is intertwined with the invention of a moral code to turn the global imagination of "made-in China" and "copycat China" into "innovation China." A copycat (*shanzhai* 山寨) means a parody and botched imitation of foreign brands. During the early period of export-oriented industrialisation in the 1990s and 2000s, *shanzhai* products were seen as representations of China's low-end production and innovation-starved system. Currently, however, production using copycat systems in Shenzhen and the Pearl River Delta has become a positive activity to self-identified techno-hobbyists and makers.

The moral code of this new generation of innovators promotes individual preferences in tech-innovation. Furthermore, it is based on their desire to break the monopoly of large European and American companies on information technology and use open-source software "to provide individuals with more diverse and more interesting user experiences of the product."[8] Many techno-hobbyists justify their activities by saying that they want to design and produce copies modifying the appearance of authentic products, and they position this practice as innovation.[9] In order to perform as innovative subjects rather than copycat producers, these maker-entrepreneurs rebrand "copycat China" into "innovation China" by giving copycat production a positive meaning. "Products always consisted of hacker technology and disruptive innovation to renovate Western-designed manufacturing products. The ambition of making copycat is what I call the spirit of Makers." This was stated by an industrial designer, Xiao Bo, who wants to sell copycat smart bracelets.[10] Forerunning Chinese makers overseas and some foreign makers also join in the reproduction and promotion of this idea: "copycat is the innovation in China" (Liao 2017). In their mindset, in the outsourcing system of the global economy, intellectual property (IP) is a tool to maintain the high-value position of "Western designers" and the low-value position of "eastern manual laborers." "This is unequal, that the Westerners have owned the discursive power for a long time," said Xiao. "Why not Chinese *shanzhai* as an innovation?"

The new inventions of people like Xiao include varied electronics: from smart phones to customised devices such as power banks, consumer-end robotics, and health-data calculators, which can affiliate with Apple or Samsung smart phones using platforms offered by companies like Tencent, Baidu, and Alibaba. Most entrepreneurs making smart devices in Shenzhen are like Xiao in the firm belief that their cottage industry has a certain degree of innovation. The goods sold in the electronics market have "applied new information technology innovations for upgrading people's lives. IT made the ordinary assembly line product have more intelligence for human life."[11] This mindset fits well with the central state's policy that manufacturing goods are to be upgraded with the assistance of information technology to make a smart city and city of innovation.

## Constructing Incubators for a Smart City: The Association of Big Enterprises and Small Start-ups

In addition to reinventing markets and attracting IT professionals, fostering an incubation system is critical for the local government to facilitate Shenzhen's transformation into a city of innovation and smart city. Since Shenzhen's government has been promoting innovation, entrepreneurship, and smart city policies, the city's number of smart hardware start-ups has surged (Shenzhen Bureau of Commence 2021). The incubator system for these start-ups imitates the Silicon Valley model. For example, the business incubator provides professional services such as consulting, patent registration, and business registration for entrepreneurs. The Silicon Valley model is characterised by an industry-academia-research economy centred on universities or research institutions and supported by financial institutions (Aoki 2000). Most start-ups active in Shenzhen's hardware innovation sector have emerged through incubators and start-up competitions held by innovation associations. However, the incubator system in Shenzhen is also different from the Silicon Valley model in several ways.

First, the rise of incubators in Shenzhen is based on industrial upgrading and urban renewal planned by sectors of the local government (e.g., Tech-innovation Bureau and Shenzhen City Planning and Land Resources Committee) and the local National Development and Reform Commission (local NDRC). Since urban land in China is owned by the state, local governments convert abandoned factories into industrial parks to house innovative incubators. Through activities such as start-up competitions organised by the local government, entrepreneurs receive support from the local government in the early stages of their business. Most entrepreneurs who register their start-ups in Shenzhen can rent space at the incubator for less than average market rent, or even use the incubator's office space for free. At the same time, urban industrial sites are often redeveloped into high-end residential areas to attract financially established talent. Shenzhen's government has planned talent housing (*rencai gongyu*) in different areas, and technology entrepreneurs who meet the criteria for talent recognition in Shenzhen can apply for talent housing at a low price.

Second, most of the incubators in Shenzhen are registered as private non-enterprises.[12] Under the national Mass Entrepreneurship and Innovation plan, business incubation is seen as a public service. By providing advice to entrepreneurs and a network of investors, incubators registered as private non-enterprises are able to take on public services from the local government. This takes the form of the government procurement service (*zhengfu goumai fuwu*) framework, a Chinese version of public-private-partnership (PPP). It means that the local government can outsource their duty of promoting entrepreneurship and innovation to non-governmental organisations. In some sense, the local government distributes accountability, as well as policy risk, to the organisations. Since the 2010s, the number of incubators that were registered as private non-enterprises in Shenzhen has increased significantly.[13] It is worth noting that the emergence of private non-enterprises does not mean the growth of civil society in Shenzhen. Most of incubators were founded on properties owned by either local government sectors or private real estate enterprises. According to an interviewee's explanation, this organisational change "is to allow the government to adapt to market-oriented needs and to foster an intermediary institution that can interface with entrepreneurs, investors, and the market."[14]

Third, Shenzhen's venture capital funds mostly come from the local governments and large local enterprises. As a sub-provincial city with little local debt, Shenzhen used fiscal surplus funds to create a municipal government-guided fund with a total size of more than 100 billion RMB in 2015. It is one of the earliest and largest government-guided funds in China. An interviewee explained that the local government wanted to promote competition in the financial market and participate in the international venture capital market to attract more investors, so the fund manager separated ten billion from the government-guided fund and established Shenzhen Angel FOF Management Co., Ltd. Shenzhen Angel FOF is the largest government-guided fund for investment in China. Tencent, Huawei, ZTE, Ping An Group, and Vanke, which are flagship private companies in Shenzhen, have also been encouraged to become investors in rising start-ups by the local government. It should be highlighted that most of the large Chinese tech-companies active in national and international markets are private companies, but they maintain a relationship as well as tension with Shenzhen's local government. On the one hand, they are tax generators and thus favoured by local industrial policy; on the other hand, they are subject to the policy decisions of the local government., They maintain a certain cooperative relationship with each other, which I will expand on in the next section.

To sum up, the local government in Shenzhen SEZ has had, and continues to have, a salient role in establishing an incubation system. Not only does it guide large local enterprises into the field of innovation and entrepreneurship investment, but it is also an investor itself. This reflects the more generally changing relationship between the Chinese local state and local enterprises: in the process of promoting entrepreneurship and innovation, local government is constantly mobilising local enterprises and social associations to take on economic and social management tasks that cannot be accomplished through administrative force alone.

## Shenzhen's Innovation as an Expanded Ecosystem Model in China: The Reassemblage of Local Governments and Local Enterprises?

The result of Shenzhen's practice of urban and innovation policy is the Smart Shenzhen project. This reflects the uniqueness of the city's political and economic transformation as a special economic zone: the industrial solutions that were tested and proved viable in Shenzhen were adopted by

Beijing and then replicated elsewhere. In this way, Shenzhen has become a model to be followed (O'Donnell, Wong, and Bach 2017). China studies and urban anthropology have done much to analyse the political economy of this Chinese model city. Modelling is, overall, a common way of governing urban areas in contemporary, late-socialist China (Hoffman 2010). However, when looking at Smart Shenzhen as a model, we have to realise that the institutional phenomenon of modelling is rooted in the tension between central and local government in China. This tension reverses the industrial blueprint of Smart Shenzhen drawn by Shenzhen's own government.

The urban-planning framework of Smart Shenzhen originally used Silicon Valley, the EU's smart city framework, and Singapore's Smart State as references (Shenzhen STITITC 2013). Although this framework emphasizes the Ministry of Science and Technology's launch of the smart city theme in the national 863 program[15] at the end of 2010, Shenzhen has become a typical representative of China's Silicon Valley. Moreover, Smart Shenzhen is seen by pragmatic local government officials as a model for transforming government services through industrial transformation. An expert who has participated in Shenzhen's smart city planning told me that they did not choose smart city as a brand at the beginning. The momentum of Shenzhen's industrial renewal made the government realise that using the name Smart Shenzhen to justify the city's role as an innovation leader would help launch other industrial upgrading projects. My interviewee said, "Around 2010, some new technologies and technical terms began to appear in the industry, such as Internet of Things (IoT) technology, which is a framework for understanding cities as physical entities," continuing that "The IoT was first applied to solve the problem of manual water meter reading in hydropower stations: by installing sensors on domestic or public sluices, hydropower stations can account for household and even city water consumption with relative accuracy."[16] Beyond solving the specific problem of reading water meters, the Smart Shenzhen industrial plan aims to change the city management model, to improve management efficiency and provide a strong technical guarantee. According to the policy discourse, its purpose is "to enhance the monitoring, analysis, early warning, decision-making capacity and wisdom of urban management (Shenzhen STITITC 2013)." So far, this IT-driven smart industry is regarded by local government as a source of technical social management tools for public institutions.

However, the government's hopes and expectations for smart industries have not materialised smoothly. The plans of industrial policy experts to solve management problems with technology require the cooperation of various departments within the government. Smart Shenzhen requires various government sectors and local companies to integrate their data on one platform. This data centring is not supported by all government departments nor by most enterprises and, so far, interests are simply not aligned.[17] This situation is echoed by mainstream research on smart industries and smart city construction in China (Große-Bley and Kostka 2021).

Based on this finding, I further my argument that, when it comes to actualising Smart Shenzhen only the industrial transformation aspect is a relatively smooth process. Shenzhen's large companies in the information industry, such as Huawei and ZTE, have become world-renowned smart city hardware and software suppliers. For example, through the smart city concept and 5G base station construction, Huawei has launched Huawei Cloud, Government Cloud, Kunpeng, and other products. Smart Shenzhen has been concretised by Huawei, ZTE, Tencent, and Ping An Insurance as a business template for urban renewal. Ping An Group, which has a full financial license, started to build Shenzhen's government services app, i-Shenzhen, in January 2019. It covers services and information regarding social security, health care, transportation, police security, life insurance, cultural, sporting events, and other areas. Moreover, these large private enterprises are driving the transformation of an outsourcing chain of IT service products. Shenzhen's hardware manufacturers (those copycat hardware suppliers who were original equipment manufacturers for European, Japanese, and American companies), software application startups, and small and medium-sized enterprises in Shanghai and Ningbo provide technical support for Huawei's Kunpeng server board.

In addition to serving to strengthen the local government's administrative and economic legitimacy, the smart city branding is a tool for local companies in Shenzhen to expand and create impact beyond the city's geographic boundaries. The largest client base for IT companies expanding their smart city templates outside Shenzhen is made up of the governments of other Chinese cities. Various local governments are moving through the Shenzhen experience, constantly competing for financial policy support from Beijing to build new smart city infrastructure. Thus, Shenzhen's information industry has become contractors for other local city governments in China. These governments try to compete to create the label of an advanced smart city by using service offerings from IT companies in Shenzhen with sophisticated smart city plans, thus further attracting investment. As explained by one interviewee, when many local governments order this kind of product and think it is useful, others will imitate this behavior so as not to be left behind. Of course, Shenzhen IT companies themselves have done a lot of marketing this end. More than 120 Chinese cities, such as Shenzhen, Shanghai, Ningbo, Weifang, Yiyang, and Dunhuang have purchased Huawei's smart city services. However, it is worth noting here that after purchasing the smart city services, each local government mainly aims to use IT enterprises to drive the innovation of local industries. For example, the Ningbo government's 14th Five-Year Plan for Smart City Construction, launched in 2021, highlights that the main purpose of Ningbo's smart city planning is to promote "industrial integration and digitally driven industrial development" (Ningbo Big Data Bureau 2021).

Shenzhen's smart city template was also affected during COVID-19. Local government and business relationships have undergone several major changes in the process of building Smart Shenzhen through the pandemic; these changes are best considered in terms of how the local government and residents are relooking at prevention and governance, the relationship between local and central governments, and the relationship between the state and the market. As COVID-19 proliferated, Shenzhen took the lead in launching a Health Code platform, designed and provided by Tencent. Leveraging the strong interpersonal network reach of the company's popular WeChat, China's largest social networking platform, Health Code became the most heavily used platform during the pandemic. i-Shenzhen, the government app promoted by Ping An Group in 2019, did not have many registrants initially, but the number skyrocketed during the pandemic. And, following Ping An's initiative, Huawei and Tencent formed a strategic partnership at the end of 2020 to help some district governments in Shenzhen build smart city projects; Tencent's Health Code and i-Shenzhen services can be embedded with each other. So far, the provincial and municipal government-led smart city framework seems to have been replaced by a polycentric, district-based design. The government market for smart cities, in turn, has taken on a multipolar monopoly pattern. But fears of a monopolies on technology platforms have also increased as the pandemic continues in early 2022. It will take time to see if Smart cities will move beyond the Shenzhen model into re-assemblages of local governments and enterprises in a polycentric state.

## CONCLUSION

The sociological exploration and explanation of the Shenzhen SEZ's transition shows a strong push toward a digitally driven innovation ecosystem in China. The Shenzhen SEZ, the former borderland of the market economy under socialism, has been de-bordered and standardised into a City of Innovation and smart city driven by local government, migrant young professional, and IT-intensive enterprises. This chapter finds that the production of this innovation ecosystem is a process of transforming market space, upgrading industry, and increasing the mobility of people and capital.

First, this chapter has analysed how young tech-professionals experience the production of this City of Innovation: how they remake working patterns and their subjectivities to fit into the supply chain of the new urban economy. This itself is a key objective of local governance in Shenzhen.

Shenzhen's plan to establish itself as a City of Innovation and its implementation of Mass Entrepreneurship and Innovation policy aims to attract talented people with IT skills. The hope is to foster indigenous innovation projects, to meet the growing needs of the domestic market, and to further replace the export-oriented economic model of early industrialisation. Second, the chapter has shown that Shenzhen's innovation agenda entails a particular process of formalising the informal economy, a process driven by the local authorities. This was pioneered by the local government in Shenzhen, as the city was the first SEZ of China's reform and opening up. In this context, formalisation means that the local government encourages and even creates grassroots entrepreneurial activities with the support of large local enterprises (e.g., IT companies, real estate companies, and financial institutions) and NGOs. This further attracts talented to Shenzhen. Third, in the tension between the local government and the central government's macroeconomic policy, the local government has adapted its Smart Shenzhen project under the aegis of the national smart city industrial policy. In practice, because the application of smart city policy has not inherently changed the governance paradigm in China, the local government sees the urban planning of Smart Shenzhen primarily to promote industrial upgrading. This local strategy has ultimately strengthened the collaboration between the local government and local enterprises and start-ups, further expanding Smart Shenzhen as an innovative model for other cities in China.

## NOTES

1. Three Import and Compensation Trade Enterprises (*sanlai yibu*): shorthand for enterprises that process imported raw materials, manufacture products according to imported samples, assemble imported parts, and repay loans for imported equipment and technology with products. Emerging along the coast in the late 1980s, all these enterprises export their products abroad.
2. Interview Huang, 12 September 2012.
3. Interview with Fu, an architect working at the Institute for Rural and Urban Planning Shenzhen, 5 October 2015.
4. Interview with Huang, 19 October 2015.
5. For more information about the influence of this policy purge, see "The Cleansing of Copycat Cellphones," available at: http://finance.ifeng.com/news/tech/20130103/7507386.shtml (Accessed 8 August, 2018).
6. Interview with Cui, 22 November 2015.
7. Mass Innovation and Entrepreneurship is an innovation and entrepreneurship policy promoted by the State Council of the People's Republic of China since 2015. This policy advocates sub-national government to promote entrepreneurship among professionals, especially young people, in the technology sector. This arrived in the

context of the rapid growth of the new global economy and the rise of internet technology. Following this policy, local governments have issued policy documents to support the construction of tech-entrepreneurship infrastructure (e.g., crowdsourcing spaces, entrepreneurship guidance funds, and urban redevelopment funds).

8. Interview with Xiao Bo, 6 December 2015.

9. Interview with Xiao Bo, 6 December 2015.

10. Interview with Xiao Bo, 6 December 2015.

11. Interview with Bai, 22 November 2015.

12. According to the Interim Regulations on Registration and Administration of Private Non-enterprise Units, the phrase private non-enterprise refers to social organisations and other social forces, as well as private citizens, using nonstate assets to engage in non-profit social service activities.

13. For more information about the influence of this policy purge, see "Hundreds of Innovation Incubators: How Can the Quantity be High Quality?" available at: http://finance.china.com.cn/roll/20150727/3252851.shtml (Accessed 8 August 2018).

14. Interview Cai, 8 February 2016.

15. The 863 Program was approved by the State Council in March 1986 to promote the development of high technology in China. This program began with an emphasis on government policies and funding to nurture scientific and technological talent and to support research in basic subject areas. Since then, as local governments have worked towards and reshaped this central macro-goal, the central government has continually revised the goals of the program. The most recent goal proposed that China should not imitate the West, but rather focuses on "indigenous innovation." The program ended in 2016. For a more in-depth discussion of its evolution, please read: Zhi, Q., and Pearson, M. M. 2017. China's hybrid adaptive bureaucracy: The case of the 863 program for science and technology. *Governance* 30(3), 407–24.

16. Interview with Zhang, 20 June 2022.

17. Interview with Zhang, 20 June 2022.

## REFERENCES


Aoki, M. 2000. Information and governance in the Silicon Valley model. In X. Vives, ed., *Corporate Governance: Theoretical and Empirical Perspectives*, New York: Cambridge University Press, 9–20.

Große-Bley, J., & Kostka, G. 2021. Big data dreams and reality in Shenzhen: an investigation of smart city implementation in China. *Big Data & Society* 8(2).

Hoffman, L. (2011). Urban modeling and contemporary technologies of city-building in China: The production of regimes of green urbanisms. In A. Roy and A. Ong, ed., *Worlding Cities: Asian Experiments and the Art of Being Global*, Hoboken: John Wiley & Sons, 55–76.

Huang, P. C. 2009. China's neglected informal economy: reality and theory. *Modern China* 35(4), 405–38.

Hu, A. G., and Zhao, L. 2006. Zhongguo chengshihua zhuanxin zhongd feizhengshi jingji yu feizhengshi guyong (Informal employment and informal economy in the

economic transformation in the process of urbanization in China 1990–2004). *Qinghua daxue xuebao* (Journal of Tsinghua University) 3(21), 111–19.
Jiang and Xuan. 2018. Shenzhen 36sui erxianguan zhengshi xiaxian, ceng jianzheng Shenzhen gaige kaifang (Shenzhen 36-year-old "second line off" officially offline had witnessed the reform and opening up of Shenzhen). *People.cn*. 16 January. Avaiblable from: http://politics.people.com.cn/n1/2018/0116/c1001-29766843.html [10 March 2018].
Liao, Qi. 2017. Zhongguo chuangke jiaofu Li Dawei: Shenzhen chuangchengle guigu wenhua, shanzhai jiushi chuangxin (The godfather of Chinese maker David Li: Shenzhen has inherited the culture of Silicon Valley. Copycat is innovation). *Yicai.com*. 3 November. Available from: https://www.yicai.com/news/5364934.html [10 March 2018].
Lindtner, S., Greenspan, A., and Li, D. 2015. Designed in Shenzhen: shanzhai manufacturing and maker entrepreneurs. In *Proceedings of The Fifth Decennial Aarhus Conference on Critical Alternatives*. Aarhus: Aarhus University Press.
Mathews, G. 2011. *Ghetto at the Center of the World: Chungking Mansions*, Hong Kong: University of Chicago Press.
Ningbo Big Data Bureau, Economic and Information Agency, City Development and Reform Commission. 2021. Guanyu yinfa ningboshi zhihui chengshi jianshe shisiwu guihua de tongzhi (Notice on the issuance of the "Ningbo smart city construction" in the 14th Five-Year Plan for economic and social development), issued 13 July. Available from: http://www.ningbo.gov.cn/art/2021/7/15/art_1229095999_1652142.html [22 June 2022].
O'Donnell, M. A., Wong, W., and Bach, J. eds. 2017. *Learning From Shenzhen: China's Post-Mao Experiment from Special Zone to Model City*. Chicago: University of Chicago Press.
Pieke, F. N. 2009. *The Good Communist: Elite Training and State Building in Today's China*. Cambridge: Cambridge University Press.
Pu, Defa and Li Bin. 2003. Shenzhen chengshihua tisu, jiang chengwei quanguo shouge wu nongcun chengshi (The acceleration of Shenzhen's urbanization: Shenzhen will be the first country-less City in China). *China.com*. 8 August. Available from: http://www.china.com.cn/chinese/2003/Oct/432834.htm.
Shenzhen Science, Technology, Industry, Trade and Information Technology Commission (Shenzhen STITITC). 2013. Zhihui Shenzhen guihua gangyao (2011–2020) (Smart Shenzhen Planning Outline), issued 22 April. Available from: http://www.szns.gov.cn/xxgk/qzfxxgkml/ghjh/zxgh/content/post_3708896.html [12 June 2017].
Shenzhen Bureau of Commence. 2021. Shenzhen rengong zhineng jiqunhua fazhan, qiye shuliang weiju quanguo dier (Shenzhen Artificial Intelligence Cluster Development, the Number of Enterprises Ranked Second in the Country), issued 26 July. Available from: http://www.sz.gov.cn/cn/zjsz/fwts_1_3/yxhjjc/content/post_9019869.html [17 June 2021].
State Council. 2010. Guowuyuan bangongting guanyu yinfa daji qinfan zhishi chanquan he zhishou jiamao weilie shangpin zhuanxiang xingdong fang'an de tongzhi (General Office of the State Council on the Issuance of the Special Action Plan

to Combat Infringement of Intellectual Property Rights and Sale of Counterfeit Goods), issued 05 November. Available from: http://www.gov.cn/zwgk/2010-11/05/content_1739089.htm [12 June 2017].

State Council. 2015. Guowuyuan guanyu dali tuijin dazhong chuangye wanzhong chuangxin ruogan zhengce cuoshi de yijian (Opinions of the State Council on Several Policies and Measures for Vigorously Advancing the Popular Entrepreneurship and Innovation), issued 11 June. Available from: http://www.gov.cn/zhengce/content/2015-06/16/content_9855.htm [12 June 2017].

The Smart City Research Institute of China Electronics Technology Group Corporation (The Smart City Research Institute of CETGC). 2017. Shenzhen xinxing zhihui chengshi sheji linian yu Shijian (Shenzhen New Smart City Design Concept and Practice), issued 17 May. Available from: https://unece.org/fileadmin/DAM/ceci/documents/2017/PPP/Forum/Presentations/Qiuxia_Liang-Shenzhen_New-Type_Smart_City_Design_Concept_and_Practice.pdf [12 June 2017].

Wang, J. and Tan, Y. 2020. Social Factory as Prosaic State Space: Redefining Labour in China's Mass innovation/mass Entrepreneurship Campaign. *Environment and Planning A: Economy and Space* 52(3), 510–31.

Yang, D. L. 2001. Rationalizing the Chinese state: the political economy of government reform. In C. Chao and B. J. Dickson, ed., *Remaking the Chinese State: Strategies, Society, and Security*, New York: Routledge, 19–45.

# Index

# About the Editors and Contributors

**Martin Chorzempa** is a senior research fellow at the Peterson Institute for International Economics. He gained expertise in financial innovation while in Germany as a Fulbright Scholar and researcher at the Association of German Banks. He conducted research on financial liberalization in Beijing, first as a Luce Scholar at Peking University's China Center for Economic Research and then at the China Finance 40 Forum, China's leading independent think tank. In 2017, he graduated from the Harvard Kennedy School of Government with a master's in public administration in international development. He recently published *The Cashless Revolution: China's Reinvention of Money* (PublicAffairs, October 2022).

**Rogier Creemers** is a university lecturer in Modern Chinese Studies. With a background in sinology and international relations, and a Ph.D. in law, his research focuses on Chinese domestic digital technology policy, as well as China's growing importance in global digital affairs. He is the principal investigator of the NWO Vidi Project "The Smart State: Big Data, Artificial Intelligence and the Law in China." For the Leiden Asia Centre, he directs a project on China and global cybersecurity, funded by the Dutch Ministry of Foreign Affairs. He is also a cofounder of DigiChina, a joint initiative with Stanford University and New America.

**Mei Danowski** is a researcher focusing on cyber threat intelligence and East Asia political, strategic, and economic affairs. Her work has supported various US government organizations. She has also held roles in the private sector, including Microsoft, Accenture, and Verisign.

**Hunter Dorwart** is a Policy Counsel for the Future of Privacy Forum's Global Privacy team, where he analyzes new bills, laws, and regulatory actions around the world with an impact on the digital economy, personal data, and data flows. He has previously worked with the Federal Communications Commission (FCC) on global digital development issues as well as the Telecommunications Industry Association (TIA), a leading standard-setting organization for the information communication technology industry. Prior to that, he spent time with a boutique law firm and the International Bar Association (IBA) where he helped coordinate a project exploring the legal implications of artificial intelligence and next-generation technologies.

**Jamie Horsley** is a visiting lecturer in law and a senior fellow of the Paul Tsai China Center at Yale Law School. Her project work and research revolve primarily around issues of administrative law, governance, and regulatory reform, including promoting government transparency, public participation, and government accountability. She was formerly executive director of the Yale China Law Center. Prior to joining Yale, she was a partner in the international law firm of Paul, Weiss, Rifkind, Wharton & Garrison; commercial attaché in the US embassies in Beijing and Manila; vice president of Motorola International Inc.; and a consultant to the Carter Center's China Village Elections Project. She holds a BA from Stanford University, an MA in Chinese Studies from the University of Michigan, a JD from Harvard Law School, and a Diploma in Chinese Law from the University of East Asia.

**Adam Knight** is a Ph.D. candidate at the Institute for Area Studies at Leiden University, where he focuses on the design, implementation, and consequences of the Chinese social credit system. Adam holds a First-Class degree in Chinese Studies from the University of Oxford, during which time he was a Fung Scholar. Adam also holds an MSc in social science of the Internet (distinction) from the Oxford Internet Institute, where he was a Henfrey Scholar. Adam frequently contributes to reports in media such as the BBC, *Financial Times*, and South China Morning Post.

**Genia Kostka** is a professor of Chinese politics at the Freie Universität Berlin. Her research interests include digital media and technologies, digital participation, and digital governance with a regional focus on China. For her current ERC Starting Grant (2020–2025), she examines how local governments use digital technologies for urban governance in China.

**John Lee** is director of the consultancy East West Futures. He is also a researcher at the Leiden Asia Center, a consultant for the International Institute for Strategic Studies and co-lead on the EU China Semiconductor Observatory. Lee's research focuses on China and digital technology, in particular the semiconductor industry, China's cyberspace governance regime, and future telecommunications networks. Previously he was a senior analyst at the Mercator Institute for China Studies and worked at the Australian Department of Foreign Affairs and Trade and Department of Defence. John holds a master of laws from the Australian National University and a master of arts from King's College London.

**Gianluigi Negro** is an associate professor in Chinese Studies at the University of Siena and research a fellow at the China Media Observatory (CMO), Università della Svizzera italiana (USI), Switzerland. After his Ph.D. at USI, he was a post-doctoral research fellow at the School of Communication of Tsinghua and Peking Universities. He is a member of the Global Internet Governance Academic Network (Giga-Net) and serves on the international editorial board of the Palgrave Series in Asia and Pacific Studies, as well as the editorial board of *H-Hermes—Journal of Communication* and the *Journal of Transcultural Communication*. His research focuses on Chinese media history and Chinese Internet governance.

**Straton Papagianneas** is a Ph.D. candidate at the Institute for Area Studies at Leiden University, where he studies the automation and digitisation of legal courts in the People's Republic of China. He is strongly interested in the study of Chinese law and society, governance, automation of justice and administration, and its surrounding ethical and normative questions. He also teaches courses on Chinese law and society, governance, and automation and justice. Outside of his academic work, Papagianneas is an editorial board member at Sinotalks, a trusted repository of knowledge about Chinese law and policy that focuses on elucidating proposed, pilot, and polished solutions to problems related to China and discussing their global implications. Previously, he was a senior editor of *Mapping China Journal* and assistant managing editor of the China Guiding Cases Project at Stanford Law School. Straton is also affiliated with the Leiden University's VVI Institute for Law & Society, and the e-Law Centre for Law and Digital Technologies.

**Tim Rühlig** is a research fellow at the German Council on Foreign Relations (DGAP) researching Europe-China relations, Chinese foreign and industrial policy including high technology, and Hong Kong politics. His current projects focus on China's growing footprint in technical standardization, the emerging US-China technology rivalry and its implications for Europe as well as the politics of Hong Kong. He is a member of the China Task Force of the European Standardization Organizations CEN and CENELEC, and a member of the Management Committee and the Core Group of the EU-funded COST Action "Europe in China Research Network" (CHERN), where he chairs Working Group "High technology and Innovation." He holds a Ph.D. from Frankfurt University with a thesis on sovereign state control in China's foreign policy, and recently published China's Foreign Policy Contradictions (Oxford University Press, 2022)

**Yujing Tan** is a tutorial lecturer in international studies at Leiden University. She conducts ethnographic research about "doing innovation" in the socio-technological transition of China. She approaches the emergence of Chinese innovation, a political economic transition as well as a social movement, from the perspective of economic sociology and anthropology. Focusing on transformation of the Leninist state, her research includes Chinese local governance, industrial upgrading and innovation in China, and more recently China-driven internationalization.

Milton Keynes UK
Ingram Content Group UK Ltd.
UKHW011320311023
431668UK00006B/36

9 781538 184417